Mastering

Geography

Macmillan Master Series

Accounting	German
Advanced English Language	Global Information Systems
Advanced Pure Mathematics	Internet
Arabic	Italian
Banking	Italian 2
Basic Management	Java
Biology	Marketing
British Politics	Mathematics
Business Administration	Microsoft Office
Business Communication	Microsoft Windows, Novell Netware
C Programming	and UNIX
C++ Programming	Modern British History
Chemistry	Modern European History
COBOL Programming	Modern World History
Communication	Networks
Counselling Skills	Pascal and Delphi Programming
Database Design	Philosophy
Economic and Social History	Photography
Economics	Physics
Electrical Engineering	Psychology
Electronic and Electrical Calculations	Shakespeare
Electronics	Social Welfare
English Grammar	Sociology
English Language	Spanish
English Literature	Spanish 2
Fashion Styling	Statistics
French	Systems Analysis and Design
French 2	Visual Basic
Geography	World Religions

Macmillan Master Series
Series Standing Order ISBN 0-333-69343-4
(outside North America only)

You can receive future titles in this series as they are published by placing a standing order. Please contact your bookseller or, in case of difficulty, write to us at the address below with your name and address, the title of the series and the ISBN quoted above.

Customer Services Department, Macmillan Distribution Ltd
Houndmills, Basingstoke, Hampshire RG21 6XS, England

Mastering

Geography

Paul Ganderton

MACMILLAN

First published 2000 by
MACMILLAN PRESS LTD
Houndmills, Basingstoke, Hampshire RG21 6XS
and London
Companies and representatives throughout the world

ISBN 0–333–66574–0 paperback

A catalogue record for this book is available
from the British Library.

This book is printed on paper suitable for recycling and
made from fully managed and sustained forest sources.

10   9   8   7   6   5   4   3   2   1
09 08 07 06 05 04 03 02 01 00

Printed in Malaysia

Contents

Introduction

Geography has always been valued for its contribution to business and military matters as well as its use in schools and colleges. The Greeks and Romans used it to plot the locations of their cities and road systems. The Medieval explorers used maps to divide the world and gather its riches. In Victorian England public servants needed to know their way around the empire. In the twentieth century we increased our reliance on geography. Today we use geographical ideas to plan our towns and protect the environment. Its constant value makes it one of the biggest subject groups.

One of the most recent trends in geography has been in the use of case studies to describe basic principles. This has boosted awareness in geography because people can see how their own location can be analysed. At the same time, there has been a need for a text to bring together ideas and concepts that these case studies have highlighted but have not always described in detail. This book takes the opposite approach: to describe the major concepts of geography and show how they can be applied to a range of different cases. This text should be considered as a geographical toolbox. It gives the student of geography an opportunity to select from a range of analytical devices that can be used to understand a situation. In addition, the connection between and within subject areas is made clear so that we can see how the ideas fit into the larger geographical picture. Whilst the concepts mentioned are those most likely to be encountered at 'A' level and first-year undergraduate levels it is also designed to stimulate interest in the subject, provide a critical overview and to act as a reference source.

To appreciate geographical ideas you need to be able to see quickly which are relevant, grasp how they work and see how they fit into the bigger picture. This text is designed with these ideas in mind:

- Each chapter title page in Parts 2 and 3 has a diagram illustrating which other chapters are most influential in understanding the work.
- Each chapter has an overview diagram where you can see which elements are going to be discussed and how they are linked within the chapter. Subsequent chapter numbering highlights this. (The diagram also acts as a useful revision guide to which notes can be added!).
- Some concepts are especially important. They might be crucial in understanding one aspect of geography or they might be seen, in slightly different forms, in many sections. These ideas, which form the fundamental knowledge of the subject have been highlighted and shown in Tip boxes.

- At the end of each chapter there is a series of questions. Some are designed to test knowledge in the chapter, some ask for case study material to be added to test understanding and some are there to extend thought on key topics.
- This book has been prepared using a very wide range of current specialist texts. For those wishing to take any aspect further (or to get supplementary work) a bibliography is provided at the end of each chapter.

Geography is too big a subject to be confined to one level of study. Much of its enjoyment comes from being able to find explanations from a wide range of topics. The range of chapters in this book has been designed for a variety of student requirements and also to appeal to others who want to improve their understanding of this subject. To assist study the text is divided into three parts. The first, Global Geography, deals with some of the most pressing geographical issues. The opening chapter, Global Development, highlights key global questions and opens the way towards their study and discussion. Likewise, Politics and the European Union, critically describes the functioning of one of the most important institutions in the late twentieth century. A chapter dealing with planning overviews a key area of applied geography and illustrates a major theme of this book – integration. Often planning is mentioned as a footnote in a number of chapters e.g. transport, settlements, agriculture etc. Since we now see the need for an integrated approach to land use for example, it follows that our study should be in a similar format. The final chapters in this section reflect on two aspects fundamental to our work. Models are crucial to geography because they allow us a simpler way of gaining understanding. However, to gain this understanding it is necessary to see how they operate. One of the delights of geography is that it is not static. Changes are made all the time. The final chapter in this section deals with history and helps us to put ideas into sequence, see how they have evolved and to appreciate their significance to us today.

The next part outlines key ideas in human geography. Since geography is concerned with human use of the environment it follows we must see what human actions are. Topics covered in this section deal with all major uses of the landscape. Population and settlement deal with human numbers and their distribution, both crucial in current debates about living standards. Next there are four chapters highlighting the economic use of geography. Economic geography examines the way in which money can be used to determine the location of activities from mining to shopping. Transport details how we get there (and sometimes how we don't). Energy is both vital to living and a major political issue. It is also one of the most serious questions facing our future which makes its study a major element of applied geography. Agriculture is one of the oldest aspects of geography but it covers one of the most difficult areas of study: how do we feed the world today. Finally, there are two aspects of pollution and

tourism which are just making inroads into geography. Given the importance of both to our lifestyles it is fundamental that they are included.

A similar mix is seen in the third part which highlights physical geography. It starts with two chapters outlining the central processes of the
planet (earth and surface processes). This leads on to a chapter examining
soils and to one describing the theory of ecosystems and the distribution
of plants and animals. The next chapter covers one of the most prominent
issues today – conservation. More people belong to conservation groups in
the UK than for any other activity. About one person in four in the UK has
at least some passing interest whether via television or more direct action.
Finally, there is a chance to study the earth's biggest single region – the
oceans. Given that 1998 was the Year of the Ocean this makes it vital that
our knowledge matches such concern to enable us to make informed
choices about the future.

In amongst all the work there is one central idea – that geography is a
fascinating study: one which not only allows us to think more clearly about
issues facing us today but also one which gives us the tools to do it.

1.1 Global development

Subject overview: global development

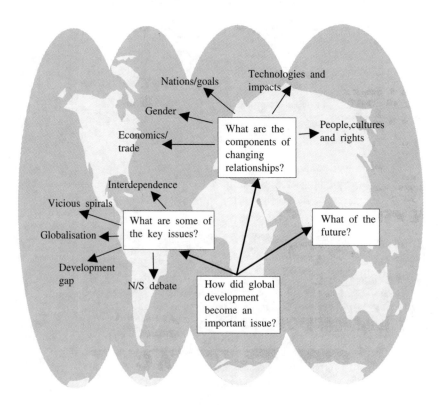

1. Global development is the idea of nations moving towards improved economic and social conditions.
2. As such, global development is a highly political and controversial subject because it deals with inequalities, how they arise and how they may be overcome.
3. Current ideas of global development started with the Bretton Woods conference in 1944 and the formation of GATT and the World Bank.
4. There are five key issues: globalisation, interdependence, development gap, North–South debate and 'vicious circles'.
5. Because of its international dimension, trade and the terms of trade are seen as crucial economic forces in global development.
6. Although global economics is important today, other perspectives are being developed e.g. gender. By examining development from a female perspective we can better appreciate many aspects of development.
7. Aid was originally part of a nation's social programme. Today, with the benefits and disbenefits better understood, aid is seen as another aspect of development. Increasingly, nations receiving aid are demanding a greater input in discussions.

1.1.1 Introduction

Global conferences such as those held in Rio de Janiero in 1992 and Istanbul in 1996 have helped to draw attention to one of the key questions facing people in general and geographers in particular: how is the world changing? To such an idea we can give the name of global development. Development usually implies a move from one set of circumstances to another, better set. In what way this is better is not always made clear but there is the idea of advancement. For the purposes of this chapter, development means a change in condition between one time to another. It does not mean that the change is either for the better or for the worse, merely that a change has occurred. We can also say that global development is all about changing relationships: between rich and poor, between North and South, between men and women. What might seem like a simple idea when you set out may become complex very easily. This chapter is all about changing perspectives – on different ways of examining the same idea and what this can tell us about how we study the world.

In any discussion of global development there comes a time when you have to group nations together to make some general statement. The trouble is that whatever is chosen there is going to be some problem of definition and meaning. Take 'Third World' for example. The very term suggests third class, less than the best etc. Compare this with the First World (USA, Europe etc.) and Second World (former communist states).

No one system is perfect and yet we need to be able to study global development. Therefore, in line with many writers the terms North and South will be used to distinguish the two areas into which this planet is often divided.

1.1.2 How did global development become an important issue?

Towards the end of World War II it was decided that there needed to be some reform of world trade (especially as it was considered to be one of the causes of the war in the first place). The Americans called a conference at Bretton Woods in New Hampshire, USA. Although the conference did not cover every issue it did give three key items for global trade: the World Bank; General Agreement on Tariffs and Trade (GATT – the main source for trade agreements); and the concept, put forward by President Roosevelt, that the world was becoming two places, developed and under-developed, and that it was in everyone's interest to reduce the gap.

Although the speech was the first official word of the **development gap** there was little done to close it. The terms of trade for goods was still in the hands of the more powerful North. The main approach to the development gap was to help with aid. By the 1970s the focus moved from aid back to the global economic system although recession in the North stopped much action. In 1980 one of the most important independent reports, 'North–South' (the report of the **Brandt Commission**) was published. Its definition and division of the world into North and South has provided a focus for the debate since then. Essentially, the North consists of the rich developed nations (but also includes Australia and New Zealand from the South) and the South are the poorer nations from South America, Africa and Asia. By dividing the world into two Brandt has helped to shape the debate. His Commission also demanded certain changes such as an end to hunger and poverty (largely through the reduction in the arms trade) and an increase in global politics. It had the positive effect of forcing people to realise that the vast majority of the world was living in poverty. Since environmental issues were global it followed that anything that happened to 'their' environment was also happening to 'ours'. A less well-discussed point was the negative side of all this. If you consider all of the South to be poor and hungry then you ignore the millions of people who are wealthy and well fed. Today, the South is taking greater control and acting with a stronger political voice which makes the development debate a very lively affair!

1.1.3 What are the key issues?

Globalisation is idea of global economic systems moving in a common system of production and trade. Take multinational companies for example. In order to make a profit they must move goods and capital to wherever they can get the best rate of return and attract the lowest costs. This way of doing business is essentially outside the national structure which may not be to any one nation's advantage. Although the basic idea is quite simple there are some important and complex implications. Firstly, if companies are globalised then so is all their production capacity. If the aim is to minimise costs then different aspects can be placed in the most suitable nations. For example, the airline industry tends to use India for data processing because it's cheap and efficient. This boosts the Indian economy (at least for those workers in jobs) but does little for other nations. The second implication is in the way the global trade expands. Japanese expansion has taken place away from Japan but in times of difficulty, returns there. What we're seeing is a core-periphery effect on trade. The core will always have the strength of the company headquarters with the periphery taking spare capacity/redundancy as required. Thirdly, it has been argued that companies are acting much like colonial nations taking only what they want from a nation and giving back little or nothing. What happens to nations that are bypassed? The fourth implication is that the government services will decline. If companies can act as if they have no home base then they can escape many of the local taxes.

Interdependence, the second issue, started as an idea in the early 1970s in Canada. The South supplied the raw materials at low cost but had to buy back more expensive finished goods. If the cost was too high they'd either buy less or borrow more which would impact on the North. This idea of trade imbalance but dependency is well known but there is another side to this. A nation tries to develop along Northern lines and borrows to build factories etc. World recession forces it into debt. How does it repay? Should it repay? The nation is dependent on finance from the North. At the same time the North's banks are dependent on getting interest on their money from the South. If that nation defaults and refuses to pay the banks are in trouble. The interdependence is the finance structure that's been created. Although power would appear to come from the North, if the South refuses to pay it actually holds all the bank's money! Some political activists have suggested that the South use its power to refuse to pay and so create banking chaos in the developed nations.

The **development gap**, the third issue, has been taken as the difference between the developing nations and the developed nations. An obvious case is the Gross National Product (or GNP – the amount of money earned by the nation divided by its total population). If some nations have a GNP of $400 per annum and another has one of $14,400 per annum

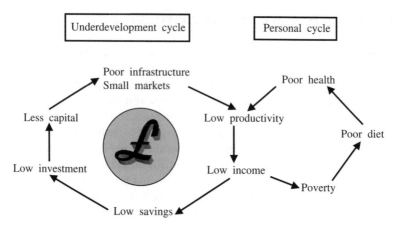

Figure 1.1.1 Vicious circles

then the difference is $14,000. Such figures don't really give an accurate picture. A better method is the Human Development Index (HDI) devised by the United Nations Development Programme. The HDI is an index made up of three different components: life expectancy, education and income. Using a series of statistical techniques every country has an HDI figure which translates into an HDI ranking. For 1995 the top three nations were Canada, USA and Japan whilst the bottom three were Mali, Sierra Leone and Niger.

Many of the newer studies are seeing development gaps not just between countries but also within countries and the gap is getting bigger. For example, if you calculate the average GNP then it does help explain the level of income but it does little to explain the distribution.

Finally in this section there is the idea of '**vicious circles**' (Figure 1.1.1). At an individual level, low income leads to poverty, poor diet, health and lower productivity (which starts the whole thing off again). At a national level low national productivity not only means fewer taxes but also less income to use to buy goods and generate wealth. Again, this spirals through the economy leaving it far weaker. It also helps to explain why food aid may not be a good thing. If you have free food then the agricultural market collapses leaving the nation with even less income.

1.1.4 What are the components of changing relationships?

In this section we can investigate various perspectives about global development. Although it starts with the most common view, economics and

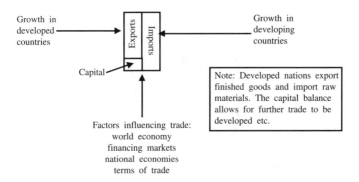

Figure 1.1.2 *Trade: an engine of growth?*

trade, that's not the only way of seeing development. These perspectives are increasingly important in human geography – it's essential to understand the world from another's viewpoint.

Many would argue that it is the **'engine of trade'** which has caused the world to develop as it has. Two theories have tried to explain this. According to the orthodox competitive efficiency models, each area exports what it grows/manufactures best. If this is carried on throughout the world only the 'best' areas will export. Because of efficiencies of production, costs will be reduced and therefore profits can increase allowing even more goods to be traded. In simple terms: trade → efficiency → income → investment → growth → trade. The other case is that of the development theorists who argue that by encouraging trade between nations, the poorer nations can export and produce a greater income. This in turn will promote the development of regions and they will in turn export and grow. Figure 1.1.2 gives some idea of the basic linkages and the key factors governing trade.

There are also spatial models to explain economic development (⇒ Rostow, economics). Barke and O'Hare's model (Figure 1.1.3) is very similar in that it assumes a progression from one form of economic development to another. People using this explanation would argue that differences between nations would be due to their different places in this model. Both Friedmann and Frank are both more spatial. They share a common idea in that trade on the coast causes a trickle-down effect into the inland areas. Friedmann assumes that this is going to create waves of development while Frank argues that it is due to penetration inland by the richer trading nation.

Even those who support these views of world development can see some criticisms. Firstly, all trade is not equal. One of the biggest problems to solve is the idea of **'terms of trade'**. This is the basis on which one nation sells something to another. This imbalance exists because raw materials are cheaper than finished goods. Thus with trade flows showing a movement of raw materials away from the Third World and a flow of

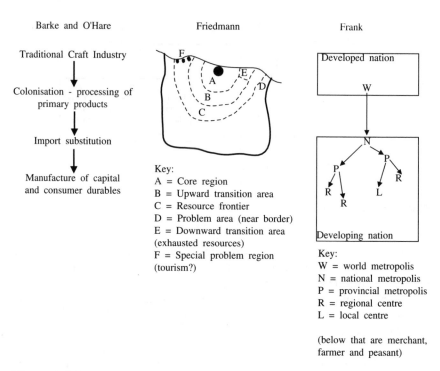

Barke and O'Hare

Traditional Craft Industry

↓

Colonisation - processing of primary products

↓

Import substitution

↓

Manufacture of capital and consumer durables

Friedmann

Key:
A = Core region
B = Upward transition area
C = Resource frontier
D = Problem area (near border)
E = Downward transition area (exhausted resources)
F = Special problem region (tourism?)

Frank

Developed nation

Developing nation

Key:
W = world metropolis
N = national metropolis
P = provincial metropolis
R = regional centre
L = local centre

(below that are merchant, farmer and peasant)

Figure 1.1.3 Development models

finished goods towards them it's not surprising that there's an unequal balance. Secondly, commodity prices will vary far more than manufactured goods will. This leaves raw material nations at a disadvantage because they can't increase the price of their produce above that which the world will pay (unless you form a cartel like the oil nations whereby the goods cannot be obtained elsewhere). Thirdly, with nations trying to improve trade with commodity prices falling there will be many nations who are in debt. The price of getting out of this would be a loan from the World Bank. Their aim would be to improve the financial situation. Often, though, this requires the receiver nation to carry out '**structural adjustment**' which is usually a term meaning reduction in public spending. Fourthly, there's the danger that by relying on exports, especially if they're primary products, you're at the mercy of global financial markets (or **trade dependency** as it's known). Finally, there's no such thing as free trade. The powerful nations will always be able to call the tune.

Gender studies, another perspective, argues that development is not a national but a gender issue. There are two lines of thought here (both of which are still being fully developed – this is a very new topic area): women's role in society and feminist theory. Take these generally accepted ideas:

- women spend far less time in school than men
- women account for more than half the hours worked but only 10% of the wages in many nations
- only one-third of women's work is in paid employment
- women have generally the key role in children's development.

It means that there's considerable inequality in the world and a great deal of it is based purely on gender. Traditionally, changes in the world's development too rarely help women. People studying this would argue that this is because women have too little power to change things although over the last 25 years this has changed slightly. Those who look at women's roles in society say that for development to be effective it needs to include all the people involved and not just the men. Because in many societies women do the majority of the work, improvement can best be made by giving women a greater role in change. One example of this is in education. Studies by the World Bank and the United Nations Development Programme have shown that female education is by far the most important aspect when it comes to improving matters. If women are given equal education then this means that their access to domestic and economic matters is improved vastly (see Figure 1.1.4). The simple ability to read means that they can understand medicine and family planning instructions. By being able to read they can also take a greater part in financial transactions – they can read and sign contracts for example.

The feminist position in geography is closely linked to the issues raised above but tends to take a more theoretical perspective arguing that the viewpoint of women is different to that of men. Take ecology as one example. In attempting to account for women's actions in 'ecopolitics' some feminist writers have stated that there are three aspects to be examined: science, rights and responsibilities, and political action. In terms of science, they argue that women's role in society gives them not only a different view but one more fundamental to the family: women's position is based on their roles as producers (e.g. paid employment) and household

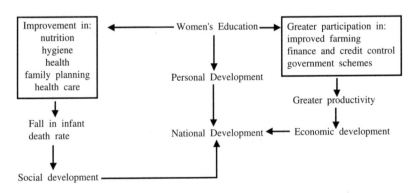

Figure 1.1.4 Women and development

organisers (i.e. unpaid employment); their central role in providing food and shelter giving them a unique view on issues such as health and environment and the idea that because of their economic and domestic work they are better placed to appreciate the holistic nature of the environment.

Similar lines of argument can be supported if you look at the feminist view on rights and responsibilities. Often (and particularly in the South) the men have the power yet the women have to take the responsibility for providing meals, shelter etc. That much of this work is unpaid gives them even less power (where economic strength is seen as power). Also since much of this work is in addition to a range of economic duties such as agriculture means that most women (again especially in the South) are working 'double shifts' making their own health less certain. Finally, in terms of political action women are often in the majority at 'grassroots' level i.e. they do the organising and campaigning at the basic level. Given that this is where most environmental campaigns are won or lost this makes their contribution especially effective.

The third perspective, **aid** (see Figure 1.1.5), tries to divide the world into two but along the lines of donor and recipient. As with gender and trade there is a power issue with the donors being the more powerful. The aid perspective often deals with some of the most difficult cases of lack of resources. Aid was set up originally to help those in greatest need from famine, disease, war etc. The aim was to provide resources where these were lacking. Today, with the demands on aid still increasing there is a tendency to see aid as a hopeless case. However, although it doesn't get the same publicity as other cases, aid has provided a whole range of success stories. A number of former aid receivers have become very powerful nations in the own right e.g. Southeast Asia nations are now better economically than their former donor nations. Millions of children have been given enough food aid to survive and diseases that cost only a few pence to treat have been reduced or even wiped out in some areas (smallpox is a good case). Unfortunately it's not just as simple as giving money. There are often strings attached. It might be that the receiver has to buy donor nation goods with the money or carry out some economic reforms that might leave people worse off. Some people in need live in countries where the rulers are not politically popular with donor nations and so no money gets through.

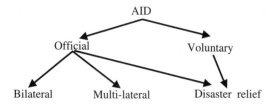

Figure 1.1.5 Divisions in aid

Technology is the fourth perspective. Those nations with the most advanced technology in any one area tend to have the greatest economic advantage. So, with this perspective, global development is seen as the way in which technology is distributed. There are two issues here: technology transfer and appropriate technology.

Technology transfer deals with the movement of ideas and machinery. Say a company has a good way of producing goods but its labour costs are too high. By moving its machinery to a country with lower labour costs it might be possible to sell more goods and make more profit. This is good for the company and for the country where the company pays its taxes but less successful for the receiver nation. The receiver nation might have the new equipment but not know-how to use it so it can gain no benefit. **Appropriate technology** has to do with the type of ideas and machinery given to a nation. In the early days of aid it was common for donor nations to give sophisticated machinery to solve Southern nation problems. The trouble was that as soon as anything broke down it was useless. Parts were expensive and workers were not trained. This led to the idea of appropriate technology – giving machinery that will actually solve the problem, that the people want and that they can operate. An early example was the egg box. Some African farmers knew that eggs would be broken easily on the way to market. The original aid solution was a high technology egg crate plant that was very efficient but that didn't employ many people and that soon broke down. Here was a solution creating more problems: underemployment was a local issue and the new plant didn't solve it and the equipment was too sophisticated for the task. The appropriate solution was to use local labour to create paper pulp which was then forced into a mould and dried in the sun. As it was labour intensive it helped local employment and because it was hand-powered it had very few moving parts, didn't need electricity, and could be repaired locally. Despite concerns about it being 'second class' appropriate technology has proved that it can work.

The final perspective examines **peoples, cultures and rights** – global diversity and personal freedom. The history of global development has been tied up with military conquest and colonialism since the fifteenth century. European expansion has rarely, if ever, been of benefit to the receiving nation: there have been numerous cases of the native people (usually called indigenous peoples) being killed and their lands, possessions etc. being taken. Today, both in the USA and Australia indigenous people win land rights i.e. they are given back land that was traditionally theirs!

By dividing the world according to peoples and cultures you get another perspective on global development. Here we see a world based not on economic power but on a range of more locally based groups. Leaving aside the moral and ethical issues there's much to be said for this viewpoint. Firstly, many groups have intimate knowledge of their local ecology and the value that it has. Often, as in the case of the Australian Aborigines, local ecological knowledge is invaluable if sustainable development is to be achieved (they've had 40,000 years of experience in Australia!). Secondly, every culture has a range of experiences that can be useful for both that groups and for others. By failing to recognise tribal boundaries the colonial settlers of Africa have created a number of difficulties that are still with us.

1.1.5   What of the future?

From the series taken above you can see that most of the answers about global development stem from wealth and power. What we've really been studying is the global distribution of power and resources. Some nations e.g. the USA have power through natural resources; others, like Japan have a very small resource base but have built up considerable wealth. How are things going to change?

There are enough resources and money to settle the majority of the most urgent problems. In fact it would only need about 10% of the global arms budget to solve problems like hunger. Sadly, the political will doesn't appear to be there at the moment. Although we could expect an increase in problems over the next 15–20 years such as:

- a growing development gap
- few ways for the majority world (i.e. South) to help itself due to lack of power
- the demand for resources will exceed the capacity to supply them
- there is the potential for trade inequality to develop further

there is no guarantee that it will happen. One of the most positive developments of the last few years is the concepts of interdependence and sustainability. Both relate to the way in which we use our global resources. Both suggest that we are all linked in our actions. Given these and the rise in political action at the local level there is every possibility that the four ideas mentioned above could be very wide of reality.

── Questions ──

1. What are the 'hidden agendas' i.e. in-built, unconscious biases in each of the naming systems: First/Second/Third Worlds; North/South, Developed/Less developed/Least developed?
2. Using library resources such as company reports, newspaper articles, study the growth and locations of a multi-national company (e.g. Nike, Nissan etc.) and a specific sector's development (e.g. computers). Draw maps showing how your examples spread from an original base. Link this growth in countries to the profits of the company/sector. Is there a relationship? How can you explain it?

What issues have been raised in the various nations that have been involved?

3. A nation might have a per capita GNP of $12,000 but it could hide a distribution range from $120 to $25,000 on an actual basis. Alternatively, a nation with $11,000 might have a range from $9,000 to $13,000. Which nation's people are better off and why?

4. Take a local political issue with a geographical idea e.g. road building, town development. Write down how you feel about the issue. How do you think it started, what concerns you most and what would you do about it? Compare your notes with a colleague of opposite gender. How do your ideas compare? What makes the differences in approach? Can you detect a specifically female/male difference? How could issues to handled to use both approaches?

5. Find out about UK aid giving. How much do we give, who gets it and what is it spent on?

1.1.6   References

Barke, M. and O'Hare, G. (1984) *The Third World*. Oliver and Boyd.

Cole, J. (1996) *Geography of the World's Major Regions*. Routledge.

Dickenson, J. et al. (1996). *Geography of the Third World*. 2nd. edition. Routledge.

German, T. and Randel, J. (1996) *The Reality of Aid 1996*. Earthscan.

Rocheleau, D., Thomas-Slayter, B. and Wangari, E. (1996) *Feminist Political Ecology*. Routledge.

United Nations Development Programme (1995) *Human Development Report 1995*. Oxford University Press.

1.2 Politics and the European Union

Subject overview: Politics and the European Union

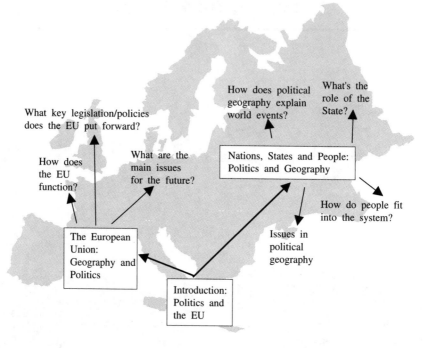

What key legislation/policies does the EU put forward?

How does political geography explain world events?

What's the role of the State?

How does the EU function?

What are the main issues for the future?

Nations, States and People: Politics and Geography

The European Union: Geography and Politics

How do people fit into the system?

Issues in political geography

Introduction: Politics and the EU

1. Today, one of the world's largest trading blocs, the EU started as a response to post-War conditions in Europe.
2. The first agreements were between France and Germany and covered agriculture and manufacturing. The central influence of France and Germany remains today and colours much of EU thinking.
3. Although seen mainly as a political unit, there are several distinct EUs: economic, political, historical, functional and physical.
4. The EU has, at its heart, a small group of officials and Ministers. The three main bodies are the Council of Ministers, European Commission and the European Parliament. With these groups, and others, legislation can be made and enforced. finance raised and key policies discussed.
5. The EU raises finance from member states. Most expenditure is concentrated on four areas: agriculture, structure, internal and external policies.
6. One of the most crucial areas is the making of policy. The most famous and controversial is the Common Agricultural Policy but there are many others. Policy-making often comes in for criticism. Although this is not always well founded, the complexity (leading to abuses) and perceived costs, do create difficulties (as does the need to work in all the national languages of the members).
7. For the near future, the main issues centre around the promotion of a more integrated Europe especially in the area of monetary union.
8. Politics is the search for, and maintenance of, power. The geography of politics shows how this can be demonstrated spatially.
9. Several models have been made to explain spatial differences. Economic theories look at the way in which trade/development has proceeded. Spatial theories examine the distribution of power blocs.
10. The state is the key feature in political geography. It defines the boundary of power and shows how power is used within the area.
11. Voting patterns show how specific political ideas are received by various groups of people. How effective this is depends on how voting is carried out.

1.2.1 Introduction

However much the study of geography involves investigation of places one simple fact cannot be ignored: most places are heavily influenced by politics. Given that the European Union is one of the world's largest power blocs it makes sense to investigate how it all started and what political forces keep it that way. This chapter seeks to examine the nature of political geography and to show its relevance to the development of a modern Europe. In doing this it divides the work into two. The first deals with the geography and ideals of the European Union: describing how it works and what it hopes to achieve. This is followed by a closer examination of the theory of political geography – the tools that can be used to analyse patterns created by the political process. Although this is a modern interpretation of political geography it should be remembered that the links between politics and geography have always been close. In the fourteenth and fifteenth centuries the major European geographers were all involved in map-making and empire-building. Since the main rivals, Portugal and Spain were neighbours it meant that maps were highly classified documents – geography was politics!

1.2.2 The European Union: geography and politics

The European Union (EU) is one of the more recent European power groupings. Its influence of every aspect of our lives is enormous. From a continent at war the EU has taken Europe towards being the largest economic force and marketplace (and yet the whole of the EU structure is run by the same number of people that would manage a large city council). Although it's a group of nations (see Figure 1.2.1) working together to promote common objectives – there's much more to it than that. It could be argued that there are several EUs each with their own characteristics, advantages and disadvantages. Firstly there's the EU as an **economic unit** where economies of scale coupled with a large, strong home market (i.e. your own nation's population) gives you an advantage when trading overseas ($\Rightarrow$ economics) because there is a greater economic base. Alternatively, the EU can be seen as a **political unit**. Like the economic

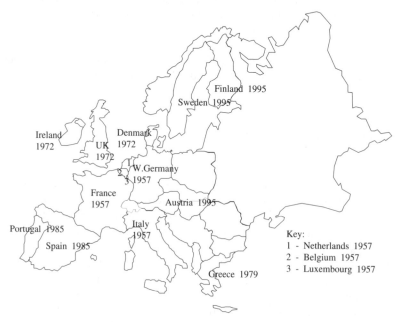

Figure 1.2.1   The European Union: member states and dates of accession

advantages, political advantages can also be gained by grouping together. It provides an answer to super-power politics and lessens the tensions that lead to war. Throughout the changes over the centuries the EU has been seen as a **historical unit**. Current arguments have tended to focus on the unity of the EU as a unit but this ignores the point that throughout European history there have been scores of boundary changes as nations/ kingdoms/empires change. It could be said that the EU is trying the impossible: to unite nations that have avoided it for centuries if not millennia. Next there's the EU as a **functional unit**. It exists because it serves a function (to reduce international tension/wars). Finally, the EU can be seen as a **physical unit**. The EU does share a common geological history. The distribution of coalfields are a good example of this as is the chalk of southern England stretching into France.

1.2.3 How does the European Union function?

The EU started as a way of escaping the political problems of the two world wars and was one of several institutions that the USA and European nations tried to start in the latter part of the 1940s. In 1950 the French foreign minister Robert Schuman devised a plan. Its aim was to promote steel and coal production between France and Germany. By 1952, when the European Coal and Steel Community (ECSC) formally came into being it had four additional members – Belgium, Italy, Luxembourg and the Netherlands. Later, after many meetings, a series of treaties was signed in Rome in 1957 (the **Treaty of Rome**). These treaties brought into being the European Economic Community and the European Atomic Community (EEC and Euratom). Nations have joined and left since that time (the original six now being 15) but the structure is still based on those three 'pillars' (i.e. ACSU, EEC and Euratom) of the early EU.

The EU is organised around three key bodies with a large number of satellites best described as a mixture of civil service and democracy (Figure 1.2.2). The **Council of Ministers** is the main law-making body. It's made up of government ministers of the 15 member states. The Council has a president (based on a sixth-monthly term, by nation). The **European Commission** is the civil service of the EU. It's headed by a president (appointed for five years) and 20 commissioners selected from member states. Its main job is to suggest legislation for the Council to debate and to enforce the legislation the Council passes. The Commission has a series of organisations working for it. Key amongst these are the Directorates General – 26 groups who take on the day-to-day running of the EU. The European Environment Agency, formed in 1990 to provide reliable data about environmental issues, is also run by the Commission. Lastly, the **European Parliament**. It's here that the Members of the European Parliament (MEP) sit. Although originally just an assembly it's now an

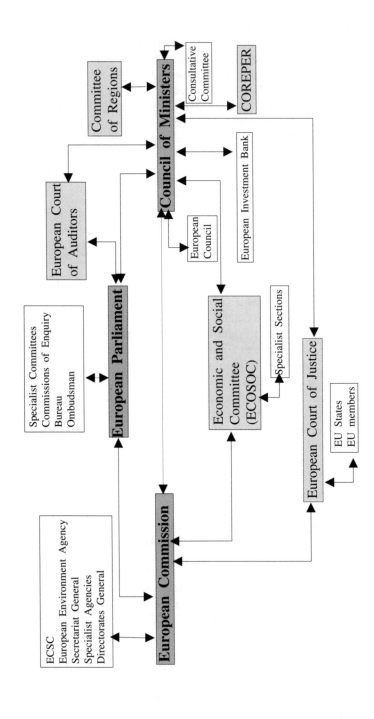

Figure 1.2.2 The structure of the European Union

elected body with a few powers. Its main job is to comment on legislation and suggest improvements. In addition to these bodies there are other minor, but crucial ones. The **Court of Justice** serves to check on legislation and those who don't keep to it. The **Court of Auditors** checks out the expenditure of the Parliament and Council of Ministers. In 1994, the **Committee of the Regions** was formed to give advice on policy matters of regional interest. Another body which expects to be consulted by both Commission and Council is the **Economic and Social Committee** (ECOSOC). It's made up of trade unions and professional organisation representatives. COREPER, the **Committee of Permanent Representatives** helps prepare work for the Council. Finally, there's the **European Investment Bank** which provides grants and loans for projects. It doesn't handle the EU money but raises it, like other banks, from investors (the member states).

There are two key aims of all the organisations of the EU: to create laws and regulations, and to enforce them. Legislation goes through a series of consultative procedures with both outside organisations and the key members of the EU (Council and Parliament). In the beginning, where a vote had to be taken, everybody had to agree (i.e. unanimous). This had been criticised as being too difficult and cumbersome so a new idea, Qualified Majority Voting was introduced in 1986. The aim was to allow for some dissent while still moving the legislation forward (one of the problems of the old system was that while waiting for everyone to agree, very little was actually passed).

EU administration is not free, neither are the programmes originally set into motion by the Treaty of Rome. In fact, at the heart of most EU work is the operation of some sort of financial control. Originally, there were four reasons for having a budget:

1 to distribute income from taxes and tariffs in approved ways
2 to support agriculture as a major sector
3 to fund administration
4 to fund a social programme.

Of course, there's only so much money that you can raise so the more the EU gets the less nations get (which makes the budget very controversial). Although some people suggest that the EU has vast resources, this is far from the truth. The EU raises money from just four sources: agricultural levies, customs, VAT and individual-nation levy. Despite the large sums involved (72 billion ECU in 1994 – about £54 billion) the actual percentage amount, at 1.24% GDP, is actually quite small. Expenditure generally falls into four categories: agriculture where the **Common Agricultural Policy** (CAP) takes about 46% of the budget; structural funds i.e. European Regional Development Fund, European Social Fund and the Cohesion Fund helping to promote regional improvements; internal research and development; and aid to specific nations e.g. the old Eastern Bloc and ex-colonies in the Africa–Caribbean–Pacific (ACP) area are favoured at this time.

1.2.4 What legislation and policies does the European Union put forward?

The EU has a reputation for producing bureaucracy and paperwork which although not always unfounded, does not give the full picture. Certain policies were at the original heart of the EU. It was recognised that only by binding legislation could the dreams of a united Europe be realised. Of course, when you start this sort of thing it can get too complex which opens it to ridicule on the one hand and fraud on the other. Although there are many areas of legitimate study to geographers this section will restrict itself to just two: the Common Agricultural Policy and Economic and Monetary Union as these best demonstrate the advantages and disadvantages of the system.

The **Common Agricultural Policy** (CAP) must be the most famous or infamous of all legislation. In the beginning, Europe had just had severe food shortages from a war which itself had followed on agricultural recession. There was a need to regulate the system and to protect both farmer and consumer. The whole aim of the CAP was to provide a stable food supply and keep the rural economy going. The key mechanism is financial. Look at Figure 1.2.3. Each year the Commission sets a **target price** for commodities (e.g. wheat). This is the price it hopes it will reach and is a compromise between farmers and consumers. At a slightly lower value is

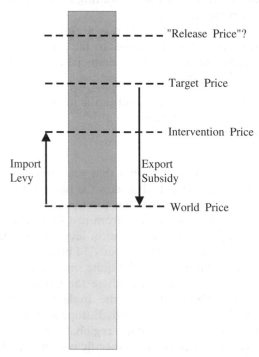

Figure 1.2.3 CAP and price structures

the **intervention price**. If there is a glut and the price falls below this the Community will buy and store surplus to artificially raise prices (giving us the wheat mountains and wine lakes of popular press). Technically, if the prices rise too far beyond the target price then stocks should be released to artificially lower prices. Anybody wanting to import into the EU must pay a **levy** if their price is lower. This is to stop dumping but it also acts as a deterrent to Third World nations. Finally, if the world price is lower than the EU price, a **subsidy** can be given to allow the farmer to compete globally.

Undoubtedly it's done a lot of good but it does have some serious problems. Firstly, it distorts the trade market and creates protectionism. Agriculture is one of the very few areas where the 'free market' is not allowed to work. By using subsidies and taxes, prices are kept higher than they would otherwise be. Secondly, as prices rise so costs get bigger. There's no end to the expenditure that could, in theory, be needed. Thirdly, the use of intervention means that food surpluses are very easily created. It costs money to store food and even more to destroy it (so it won't be sold and distort the price set-up). Finally, there is the problem of fraud. The system is so cumbersome that it's impossible for one person to know everything. This leaves it open to abuse.

The move towards economic and financial union has been part of the EU agenda since the very early days. Then it was just a vision; now it's closer to reality the difficulties can be seen. The whole process is called the **European Monetary System** (EMS – started in 1978). This was an umbrella for four separate parts of which two are most important: the Exchange Rate Mechanism (ERM) and the European Currency Unit (ECU). The ERM was necessary to stop speculation in currency dealings and to keep national currencies working within a series of limits (for inflation, interest rates etc.) The UK joined in 1990 and left in 1992 following currency speculations. The ECU was originally set up to provide a way of dealing with EU expenditure without going into national currencies. The value of the ECU is fixed against each member-state's currency which allows for exchange rates to be calculated free of the financial markets. The ECU rate is not fixed for all aspects. For example, to ensure that the CAP works evenly across different costs of living the ECU exchange rate differs – giving the idea of a Green Pound, Green Franc etc. compared with industrial and financial rates. Despite these problems, the ECU has proved a useful exercise; already some nations trade in the ECU. What of the **Economic and Monetary Union** (EMU)? Although mentioned back in 1970 it came into its own from the Single European Act in the late 1980s. Jacques Delors, then president of the European Commission set out the mechanism at the **Maastrict Conference**. Initially, all nations would have to join, and keep within, the EMS. Stage 2 started in 1994 with the aim of uniting national financial systems under the auspices of the European Monetary Institute (EMI). Providing everybody makes the convergence criteria (low inflation rate, low interest rates, modest national debt, low budget deficit and keeping within the EMS requirements etc.) it would just be possible to get a single European currency. This final stage has until 1999 to be completed. The EMS shows just how complex a seemingly simple issue can become when dealing with 15 different sovereignties.

1.2.5   What are the main issues for the future?

Events change rapidly within the evolving EU. Despite changes there are five issues which are being constantly referred to:

1 *The Single Market* – a dream since the beginning of the EEC/EU this is only slightly nearer to being a reality. What sounds like a good idea in practice is full of technical problems. How do you make sure that exchange rates actually reflect different costs? Should there be a difference in a single market? What happens if a nation can't compete? Is sovereignty lost? There are many more problemsbut it does give some idea of what needs still to be sorted out.

2 *Expansion* – at the same time the EU is trying to get 15 members to agree there are more wanting to join in. Leaving aside the problems often raised with the EU the advantages are real especially in trade and economics. To compete requires a large home market: the EU gives access to markets larger than the USA and Japan. Against this must be stressed the problems of holding together a load of nations many with conflicting ideas.

3 *Organisation* – if the market is enlarged then there's the problem of organisation. The EU can seem to be unwieldy: a vast group of administrators producing a mountain of paperwork. Despite this image, the reality is that the EU has a very small Commission of about 17,000, less than most large European cities. Most of the regulations have to be carefully phrased and translated into the 12 EU languages. If there is a problem it is that the diversity of the EU creates the need to replicate systems that wouldn't be needed in a single-culture state.

4 *Harmonisation and standardisation* – one of the main points on the way to a single market is the use of common measures and rules. Ideally, every regulation should be the same from road speed to customs paperwork. In reality this is far from the case but there is much effort going in to reduce the differences. These two terms describe what is trying to be achieved. Harmonisation covers a vast range of ideas from toys to motor vehicles. Basically, it means that everyone works to the same agreed pattern. Standardisation is very similar (and the two terms are often interchangeable) but concentrates on the agreement of set levels e.g. axle weights for road vehicles.

5 *Facing the problems* – there's been a tendency to ignore the EU until issues are really pressing. The need for a common system, so obvious

after the Second World War has declined and there are few nations where everybody from politician to voter is totally behind the EU in everything. The dreams of Monnet and Schumann, the early leaders in EU ideas, have not come to be realised in too many areas. Great things have occurred but much remains to be done. The recognition of this fact alone is vital!

1.2.6 Nations, states and people: politics and geography.

Political actions cause changes on the ground. Some political actions work at different scales or with differing amounts of power. Political institutions vary with scale – global, national, and local. The UN might be able to settle border disputes but would be completely useless with a planning application! In other words each scale has developed a structure which allows it to focus most effectively at that level. One writer, Peter Taylor takes this further by arguing that scale affects our perception. The local scale is something we experience, the national level is governed by party ideas (ideology) while at the global scale we see the reality of working in an increasingly interdependant system. Take current economic performance for example. At the local level we might experience job changes or even unemployment. At national level the political scene is set by the party in power. At a global level we have to face another set of political issues with global competition and multinational companies often setting the agenda. In another study, Immanuel Wallerstein investigated the nature of power. He divided power into four key areas which, he argued, could explain the running of the global economy: **states** (where overall control lies); **peoples** (groups with cultural links and common goals); **classes** (groups of people united by a common value e.g. religion, socio-economic level etc.); and **households** (small groups pooling their income).

1.2.7 How does political geography explain world events?

If we accept that there are political 'winners' and 'losers' then it follows that we must be dealing with an unequal distribution. In other words, inequality (whether it be in food supply, fuel, educational opportunity or

Tip

Scale matters because different things happen at different scales.

Power is about the way in which people interact and how this works out spatially.

whatever) is a political matter and can be expressed in terms of maps. The best example of this could be seen with the Brandt Report of 1987 which divided the world into two: a rich North and a poor South. The names come from geographical locations but also imply more than that. With Australia and New Zealand in the 'North' and a good deal of North Africa in the 'South' you can see that it's more of an ideological situation than anything else.

How can such global patterns in power and resources be explained? There have been numerous attempts: the best idea is to look at two contrasting ideas, **economic** and **spatial**. In terms of economics there are two ideas that need to be explored: long cycles and development take-off. Long cycles were part of the work of Kondratieff ($\Rightarrow$ economics). Another economist, Rostow devised a 'development take-off model'. (Figure 1.2.4). The five stages are:

1 a traditional society presumably agricultural with little change
2 a take-off society with new production and a powerful elite
3 take-off where there's a massive rise in production
4 drive to maturity where the whole economy improves
5 high mass-consumption with the move to consumer goods.

According to Kondratieff and Rostow we can see world political development in terms of economic changes. For Rostow in particular, only a few

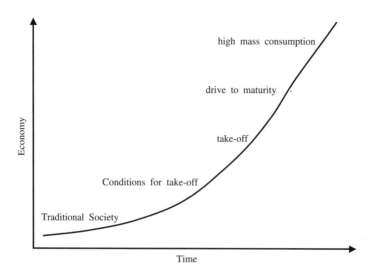

Figure 1.2.4 Rostow's model

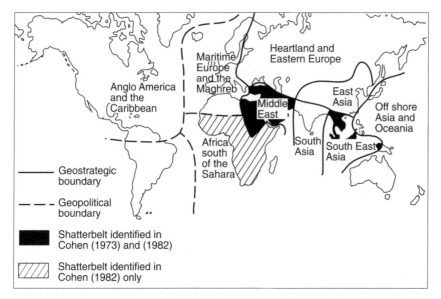

Figure 1.2.5 *Cohen's geostrategic model*

nations would break through to development at any one time. Eventually, every nation would reach stage five which he considered would be the same as the USA. The upshot of this is that poverty would occur if you didn't join in the development 'game'.

Economic models are not the only way to approach this topic. There are other, spatial, ideas that need to be examined. Much of the early work in political geography started by considering a range of global models. One of the best examples illustrating the concept is the **US World Model**. After the defeat of Germany in 1945, the USA tried to explain power groupings in terms of their distribution. At the centre was the **Heartland** i.e. Russia. As communist Russia tried to spread outwards it could only be contained by another power (i.e. the USA) which would meet it at the **Rimland**. If each 'threat' were not countered then Rimland nations would become part of the Heartland. This simplistic idea, based on a geographical notion but developed as global policy by a superpower, was with us until recently. The Rimland takeover has often been called the **Domino theory** (knock one down and they all fall). This simple model was extended by the American Saul Cohen whose **geostrategic** idea divided the world into 10 regions (see Figure 1.2.5).

1.2.8 What's the role of the state?

Geographers recognise three important divisions: state, nation and nation-state. The **state** has power over a place. A **nation** is a group of people who consider themselves to have a common history and culture. The **nation-state** is the ideal link where the state consists of just one set of people. A state is a relatively recent phenomenon linked, by many writers, to the start of the world economy in the 1500s. Before that time there were nations or even tribal areas. The state has one main function – to maintain its own security. It does this in two ways: internally and externally. Internally it keeps law and order through the justice system (i.e. police and courts) and it maintains a public view of the state (often called a belief system i.e. propaganda in support of those in power). Externally it defends its territorial interests through military and diplomatic means and its trading interests through international bodies. Although these two methods of control are common to all states the way that this is organised differs between sets of states. In effect, we can have a classification of states. Similar states would share a similar set of beliefs and controls. Figure 1.2.6 illustrates four different state models. Note that the state is trying to balance sets of forces. In democratic systems the forces are dealt with by the elected government. Imperfect states have policy determined not just by competing forces but they are heavily influenced by business and/or organised labour. Karl Marx, the influential writer whose works formed the basis of communism, considered two other versions of the state. An instrumental state (e.g. Russia before the 1917 Revolution) would have competing forces dominated by capital. A more balanced system would have labour operating against the interests of capital to produce a communist system. The world isn't standing still. There are going to be forces that try to upset the system. John Short argues that there are five forces (he calls these crises) that can upset this picture: **economic crises** occur when the ability

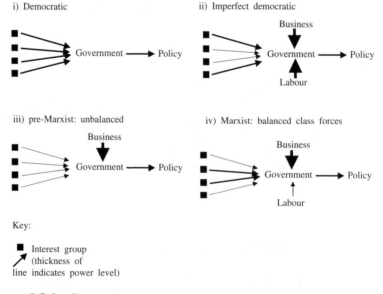

Figure 1.2.6 Power relations in the state

of the economy to deliver doesn't match the needs of the people; **legiti-mation crises** occur when the people no longer have faith in the state; **fiscal crises** occur when expenditure is greater than the money it raises; **rationality crises** occur when the policies of the state fail to deliver what they promised and **motivation crises** occur when the motives of the state are not matched by the motives of the people.

1.2.9 How do people fit into the system?

At this scale, political geography is all about how the power groups affect our lives. Because of the importance of participation in everyone's lives it is also a growth area of the subject. Rather than tackle this as one unit there are three distinct parts: voting, power structures and participation.

The **geography of voting** is, at its simplest, a map showing which party won, where. This can then be tied into other factors such as socio-economic class to provide an explanation. Models can be as simple as a map of voting (as used in the UK) to more detailed explanations of power groups and alliances. For example, a French geographer, Roccan, tied in the conflicts between political groups to the ways in which these groups allied with others with common socio-economic forces. He called this the **model of alternative alliances and oppositions**. The assumption here is that there are just two groups and each would need to form alliances to win an election. There are two other concepts that are important: representation and voting systems.

Representation is the person you get for your vote. Despite promises, all votes are not equal. Some may have a greater effect than others. Vote in a 'marginal area' i.e. where the politician has a very small majority and only a few votes are needed to remove/elect him/her. In a 'safe seat' you couldn't do it with thousands. In the UK, the Local Government Boundary Commission changes electoral boundaries to make sure that each area has approximately the same number of voters. Where the boundaries are put can be a heated debate (especially if the 'opposition' voters are put in your area). The success of boundary changes can be measured using the idea of **malapportionment** – comparing the number of voters in the smallest and largest constituencies. Thus:

$$M = L/S \times 100$$

where M is malapportionment, L is the largest and S the smallest constituency.

Unusual voting results don't have to be due to malapportionment. Another cause of unusual patterns can be due to the **voting system** you have. For the UK it's 'first past the post' but others such as preferential (a number of candidates are selected on a first, second etc. vote) and propor-

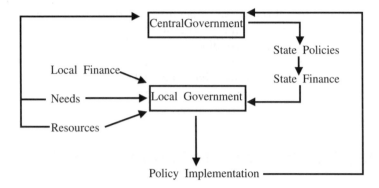

Figure 1.2.7 Centre-local relations

tional (the number of politicians are selected according to the percentage of votes their party received) systems are popular in Europe and elsewhere.

The second element is the **power structure** which can be divided into two: invisible and visible. The invisible power structure is what a government does not want its enemies (or its own people) to know about. For example, most military bases in the UK are not routinely put on maps. The visible structures are the more practically important: a good example of this is the **centre–local model** (Figure 1.2.7). There are two extremes of this model: one with a dominant centre and one with a dominant local. Because this is a power model there are always going to be political tensions. Where the balance is drawn depends on which group is stronger.

Finally, participation. **Legitimate participation** takes place through voting and through using the participation built into the power structure. What if you don't like the situation and you can't get this across? This takes us to **non-legitimate ways** (using legitimate here to mean approved channel) which means pressure groups and civil protest.

1.2.9   Issues in political geography

This final section looks briefly at some of the current questions facing political geography. They are current issues and will help to emphasise the importance of geography to our everyday lives:

Globalisation is one of the most discussed ideas at present. Although there's no real agreement over how it's defined most people would argue that it involves a move towards global structures. Thus power has moved from the tribal village to the nation to the state and now to the world. At the local level we are seeing the break up of states e.g. the former USSR and Yugoslavia into Russia, Bosnia etc. In some ways this is just a return to old political lines only on a nation rather than state basis. This has given rise to nationalism – the strong support for one's nation which can also lead to serious conflict (Bosnia/Serbia, Israel/Palestine, Russia/Ukraine to name but three). Ironically, at the same time there's been a rise in power in the centre. There are a large number of nations but with the real power in a few dominant states. A good example here is the EU with the move towards a single currency. Previously there were 15 states each with their own government and economy. Now we are seriously considering moving to 20+ nations but with only one centre.

There's more to globalisation than states and nations. One of the most important forces in recent years has been the growth of the multinational (or transnational) corporation whose loyalty is not to the state in which it was primarily based but in the company itself. It follows that it will, like any nation, make moves to defend itself. Already you can find reports of multinationals moving to/from a nation because of political/environmental etc. issues with which it disagreed. Companies such as IBM have turnovers larger than a number of Third World countries. Given this kind of power it's not surprising that it's being used. Another aspect of globalisation is equality (or equity as its often called). If power is all about winning then there must be losers. The difference between the rich and poor nations is increasing ($\Rightarrow$ global development) – the so-called development gap. If this continues then it could have serious implications. This is particularly true if you consider the rise in interdependence: meaning that we're all relying more and more on each other.

The state versus the person   To what extent does the individual have rights? The United Nations and others are still debating this. Alongside these rights we have a similar issue with the rise of nationalism (defined as having strong attachment to and support for one's nation). We can divide nationalism into three according to how we see it, the nation as: imagined community (the call for patriotism?), class struggle (in terms of workers rights); or anti-systemic struggle (such as the demand for the control of government).

Regional integration is on the rise alongside increased fragmentation. This is not the same as political union like the EU (although it is one example). Groups sharing similar goals have joined to form more powerful groups. Examples of this include: the European Free Trade Agreement (EFTA), the Council for Mutual Economic Assistance (the former communist COMECON).

Environmental concern is our final case. Ever since the first UN conference in 1972 there has been an amazing increase in environmental activity from politics to education. Because so many of the environmental problems are global ones it is reasonable to argue for another world-level force. This may well turn out to be the most important single issue in the next century.

1. To what extent is the EU working for the 'good' of the UK?
2. Using Rostow's descriptions, put in the dates when the UK reached these stages. For each of the stages, name a nation at that stage today.
3. What are the strengths and weaknesses of using Rostow's model?
4. What would be the social, economic and environmental consequences of every nation being as developed as the USA?
5. Using the EU as a case study, give examples of each of Short's 'crises' during the twentieth century.
6. What are the advantages and disadvantages of the three voting systems: 'first past the post', preferential and proportional?
7. Use newspaper reports etc. to find out about pressure groups and civil protests. Describe the groups and their activities under the headings: how, where and why were they formed; what is their organisation; how did they get resources and power; when and where did they operate; and what was the consequences of their actions? What did your examples have in common and what separated them? Why?

1.2.8   References

Archer, C. and Butler, F. (1992) *The European Community: Structure and Process*, Pinter Publishers.

Barnes, I. and Barnes, P. M. (1995) *The Enlarged European Union*, Longman.

Cole, J. and Cole, F. (1993) *The Geography of the European Union*, Routledge.

Dickenson, J. et al. (1996) *Geography of the Third World*, Routledge.

Goodman, S. F. (1996) *The European Union*, 3rd edition, Macmillan.

Knox, P. and Agnew, J. (1989) *The Geography of the World-Economy*, Arnold.

Roney, A. (1995) *EC/EU Factbook*, 4th edition, Kogan Page.

Short, J. R. (1993) *An Introduction to Political Geography*, 2nd edition. Routledge.

Taylor, P. J. (1993) *Political Geography*, Longman.

1.3 Planning

Subject overview: Planning

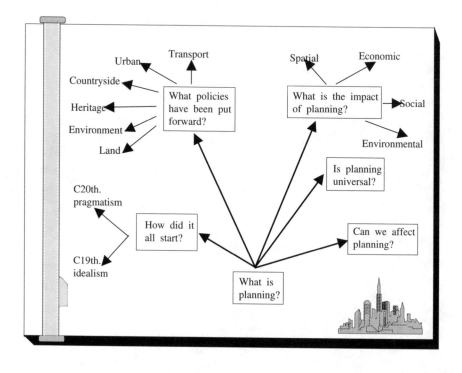

1. Planning is the control of development to balance public and private needs.
2. In the UK planning as a public activity didn't start until 1947 – it is a very recent development.
3. Planning developed from a nineteenth-century concern for social, housing and public health reforms.
4. Planning policies tend to focus on a specific issue: land use, environment, heritage, countryside, urban areas, transport and conservation.
5. In the UK there are few integrated policy plans. The best examples are structure and local plans whilst transport planning remains the least integrated.
6. Planning has four main impacts: spatial, economic, social and environmental.
7. Although each nation has its own way of planning, in Europe and North America at least, the basic principles followed are very similar.
8. Today, planning is seen as a community process – everyone should be involved.

1.3.1 Introduction

There can be few matters of geography which produce a stronger reaction from the public than planning. Although public planning in the UK is only about 50 years old it has developed to such an extent that there are very few aspects of everyday life that do not involve a planning decision. The aim of this chapter is to try to outline some of the basic principles that are involved. Often, there is a tendency to divide planning into its various categories and place each piece with the rest of the subject matter. For example, rural planning is discussed in rural development topics, urban planning in settlements and transport planning in transport. This approach has not been chosen here for three reasons. Firstly, planning involves every aspect of the community: a complete overview is necessary to understand its workings. Secondly, as projects become more complex and our understanding of impacts becomes greater so it is essential to examine all factors. This demands that all planning items are studied together. Finally, in the UK, poor decisions have often been made on the basis of limited integrated study. Transport planning is an ideal case: although there is talk of an integrated transport policy too often this means a road policy, a rail policy and an air policy etc. To truly integrate all aspects one needs the overview of all planning issues.

1.3.2 What is planning?

This must seem like an obvious question but there's more than one problem and more than one answer. A dictionary definition would argue that planning is an organised activity with a precise outcome in mind. This situation isn't helped by the fact that 'plan' suggests a document, usually involving both text and diagrams and 'planning' might contain neither. To make some sense of this, planning in this chapter is going to be used in one particular way: to describe and explain the organisation of land use changes. It's what goes on in the local council's planning department.

As was noted in the introduction, public planning is only about 50 years old. Even so, it has gone through a number of evolutionary phases. According to Peter Hall (1992), a leading planning academic, there have been three phases in the development of planning in the UK:

1 Early phase – difficult to date the start but it probably ended in the UK in the late 1940s. In this phase, planning was seen as a precise activity where master plans could be drawn up and then put into action (Hall refers to this as the blueprint era).
2 Systems phase – with the introduction of planning legislation in the late 1940s (and especially with the 1947 Town and Country Planning Act which brought the idea of local authority planning into being) there was the idea that planning could be seen as a process diagram: input, process and output.
3 Community phase – the idea that there's more to planning than a series of maps started in the late 1960s in the USA but has spread rapidly and is now a key way of working. Essentially, this would involve a more sophisticated version of the systems diagram but at most stages there would be a lot of input from local people. Thus community involvement would be seen as a major source of planning. It would be something agreed between all interested parties and not just a few key people. This might sound very good in theory but in practice it is not that easy. Most planning decisions are very minor ones usually involving little more than putting up a shed or a small extension to the house. Where there's a major decision such as a bypass or an airport terminal then there's a greater desire to get community involvement.

1.3.3 How did it all start?

The idea of planning might seem modern and the job of public planner did not come into existence until the 1947 Town and Country Planning Act but the whole idea of producing a design beforehand is ancient. One of the earliest town plans was for the Egyptian settlement of Tel-al-Amarna in the fourteenth-century BC. The street plans looked very similar to Victorian housing in industrial cities so it seems that there's very little that's totally new in planning (in terms of transport even Roman Pompeii had pedestrian, transport-free [oxen-free] zones!). So it is possible to date the start of planning to ancient times but in modern terms we really need to start with the nineteenth century.

The Industrial Revolution was a major turning point for the British economy in general and urban economy in particular. Although a few cities existed, most of the population was in small rural settlements. Thus the need for planning was virtually non-existant. However, within a century (1750–1850) the ratio of urban:rural dwellers had reversed and 75% lived in urban areas. Such a massive change created numerous problems. The lack of public investment meant that services we now take for granted were not provided. For example, it was common for water supplies to be polluted. Secondly, housing was often of poor quality and this led to both social and political unrest. Thirdly, with no control of the land market, prices often rose so much that high density housing was the only answer. So rather than expand towns to solve problems, town expansion just exacerbated them. Finally, much of the housing stock quickly became outdated and in poor condition and yet there was no effective replacement.

The key question facing Victorian town leaders was what to do about this. Although there was no legal planning structure many towns and cities had some form of planning. This was seen in three ways. Firstly, social reformers, often factory owners with high religious and moral principles, tried to solve the problem by having their own **model towns** built, e.g. Bourneville for Cadbury, Port Sunlight for Lever. Secondly, **housing reform** was carried out through a series of Parliamentary Acts to control housing quality: ventilation, room size and facilities for example. Thirdly, **public health reform** addressed the problems of the open sewers and contaminated water supplies. To provide a place for recreation and to cater for new social responsibilities many Victorian towns planned public parks and civic buildings (the start of modern urban design). Although it did much to improve the conditions for the urban poor in reality it was too small-scale and haphazard to give help to everybody.

The problems seen in Victorian towns continued into the twentieth century. With continued demand for housing the problems became worse in some cases. For example, without planning towns were prone to **urban sprawl**, i.e. unchecked growth. New ideas were tried. Many followed the Victorian ideas of housing reform (building to a better standard) and public health reform. In addition there were attempts to control what was built where, i.e. **zoning**.

One of the most influential aspects was the move towards more radical reform. This took several forms:

- Building design – Soria y Mata designed the linear city suggesting high level transportation whilst Le Corbusier tried the opposite – the vertical village. Although Soria y Mata's plans did not come into the UK we can see Le Corbusier's influence in the tower blocks much favoured in the 1950/60s.

- Urban design – a series of radical designs from the '**garden city movement**' of Howard, Unwin and Parker (giving us towns such as Letchworth and Welwyn Garden City) are still seen as highly influential. On a smaller scale, the planned 'neighbourhood' (designed by Perry in the USA) meant that social groups were seen as a key planning feature whilst the desire to contain transport led to the Radburn principle, an idea that a road hierarchy was to be devised with no major road entering a neighbourhood. Geddes and Abercrombie, working in the 1940s were concerned about the growth of London. The 1944 Greater London Plan was a grand scheme to both contain the city and allow growth away from it. From this we got not only the Green Belt but also the New Towns and Expanded Towns of Southern England.

1.3.4 What policies have been put forward?

To solve planning problems, numerous Acts of Parliament have been passed. Rather than repeat these, the aim of this section is to show how the planning system has dealt with key issues in land, environment, heritage, countryside, urban areas and transport.

Central to planning in the UK and elsewhere is the control of development. This raises an important question: who owns the land? In Britain it is assumed that somebody owns (i.e. has legal control over) every piece of land. Ultimately, planning control would take that right away. Reports proposing that all land be nationalised, i.e. be owned by the State, attracted much attention in the 1940s. How this would work, how people would take it and how (if at all) compensation would be paid created so many problems that the idea was dropped. A series of Acts from 1947 until the 1970s tried to resolve this issue with a variety of schemes from a Land Commission to a range of taxation/compensation deals. To date, no really satisfactory method has been found. Another major land issue was the containment of urban sprawl and the provision of open space for recreation. One scheme, the **Green Belt** around London tried to do both. The aim would be to stop urban growth at some line and allow it to continue beyond a second line. Another example of planning and land comes in the form of control of vacant and derelict land. This has been a strong issue because, as opponents of development on new sites (**greenfield** site development) have noted, the 'damage' from development has already been done. Often, derelict land (officially defined as land needing treatment before re-use) is found in those areas where industry has declined (bringing in social and political issues as well). These cases demonstrate some of the ways in which planning and land issues come together. There is a central issue here: what are the rights of the owners and how are these reconciled with the needs of society at large? Such questions are far from trivial – they put planning at the heart of social and political life.

Although the environment is a key topic today, in planning terms it is usually restricted to pollution and waste matters. If land was seen as a question of ownership and control then environment and planning can be seen as a question of civic duty and social concern. The London smog of 1952 was so severe that it brought about demands for the control of pollution. It was seen that only an Act would be strong enough to control the situation – the free market would be unable to help. This led to the 1956 Clean Air Act, one of the most famous modern Acts which demanded, amongst other things, the use of smokeless fuel and the introduction of smoke control areas (commonly called smokeless zones). Here we have a dilemma: how to find a course between a policy of non-intervention and demand for pollution control? This is still a question of debate.

Heritage forms a different picture. It has always been one of the clearer planning issues because the objects under consideration are easy to define. The whole issue started in the nineteenth century with the forming of the National Trust – a private charity dedicated to the preservation of historical buildings and monuments. To this day, the voluntary sector has been a major force in heritage conservation. However, the same is not true for the vast majority of buildings and remains in private ownership. Again, as with land, planning is seen as a way to protect for the public that which is private. Whether the heritage site is a building, ancient monument or archaeological site the principle is the same: conservation protection is afforded for those items meeting a specific set of criteria. In terms of monuments this would result in being placed on a list organised by the Royal Commission on Historic Monuments. The process is similar for buildings (which become known as Listed Buildings). Archaeological sites were often ignored until the 1979 Archaeological Areas Act. This created Areas of Archaeological Importance which protected remains.

Countryside planning is an example of the protection of large areas either for the enjoyment of the public or for the conservation of resources. Put simply, this means landscapes and wildlife. Both groups are subject to the notion of grading i.e. changing the rules depending on the quality of resource. For landscape areas this means National Parks, Areas of Outstanding Natural Beauty and Areas of High Landscape Value. Although there have been controversies over the years the aim of providing recreation in areas considered 'beautiful and relatively wild' (a key phrase in an initial report) still holds good. For wildlife the top places would be National Nature Reserves or Marine Nature Reserves with lower grade sites as Sites of Special Scientific Interest, Local Nature Reserves and Voluntary Nature Reserves. Protection is carried out using a variety of means: legislation to set out the ground rules, government agencies whose specialists administer and help, and private individuals who own the land.

Urban areas are amongst the most complex of all the planning issues. Rapid, unplanned growth was a feature of the Industrial Revolution: poor quality housing was the key result. This has meant that the reduction or removal of sub-standard housing has always been seen as a major element of urban planning. Slum clearance started in the nineteenth century and the 1918 Tudor Walters Report both continued it and set the standards for good housing. However, following the Second World War there was a need to address this more firmly. Widespread clearance was the first response mainly because of war damage but also as a way of clearing poor areas. Clear the area and get rid of the problem? Neighbourhoods are far more than buildings and early attempts to create new areas by just adding people failed (New Towns and tower blocks whilst excellent designs on paper were social problem areas and few were built after this became appreciated). By the late 1960s the problems were better understood and wholesale clearance was on the way out. From then on grants and small-scale schemes sought to improve specific urban areas. Thus policies had become more sophisticated moving from widespread clearance and community dispersal through general improvement through to community-oriented specific repairs and upgrading.

Given the widespread publicity to arguments over schemes such as the Newbury (Berkshire) bypass and the M3 extensions in Hampshire one could be forgiven for wondering if there is a transport policy. The history of transport is one of competing modes (i.e. methods of transport) and the resulting changes needed to be made. By the end of the nineteenth century we can say that there was the railway for long-distance transport and roads for, mainly, local travel. It was the growth of the motor car which changed the face of transport planning. The impact this had on transport planning can be seen in two 1963 reports. The first, the Beeching Report, called for the closure of numerous small lines. The other, the 'Buchanan Report' recommended the expansion of motorways and widespread traffic management in towns. Unlike any other planning area there have been no major transport Acts which integrate the various modes of transport. Since the late 1980s the impact of road schemes and the public change in attitude has meant that this remains one of the most politically sensitive planning issues.

1.3.5　What is the impact of planning?

Some idea of the extent that planning has become part of our lives can be seen in Figure 1.3.1 where all areas marked represent some aspect of planning. Unfortunately it is this very widespread nature of the impact that makes it very difficult to assess. Basically, impacts can be divided into four: spatial, economic, social and environmental.

The spatial effects depend on where you are. In town centres the commercial and business functions have been kept at the expense of housing and most town centres now have little of the social life tradition-

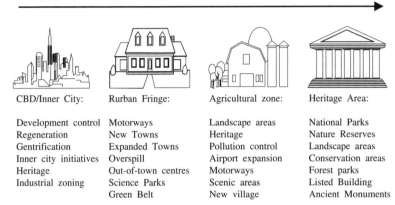

Distance from the Urban Centre

CBD/Inner City:

Development control
Regeneration
Gentrification
Inner city initiatives
Heritage
Industrial zoning

Rurban Fringe:

Motorways
New Towns
Expanded Towns
Overspill
Out-of-town centres
Science Parks
Green Belt

Agricultural zone:

Landscape areas
Heritage
Pollution control
Airport expansion
Motorways
Scenic areas
New village

Heritage Area:

National Parks
Nature Reserves
Landscape areas
Conservation areas
Forest parks
Listed Building
Ancient Monuments

Figure 1.3.1   The effect of planning

ally associated with them (leading to loss of community for example). At the same time planners have had to cope with regional and national demands where private developers have taken to using out-of-town centres for shopping. In the suburbs there has been an increasing usage of land at an increasing density putting pressure on open space and transport links. Directly outside the town there has been an increase in building as 'executive village developments' have grown up as **counterurbanisation** has increased (leaving truly rural areas suffering the effects of depopulation). At a regional level we have seen the **containment** of large urban areas which are surrounded by green belt land. Population pressures are dealt with by building new or expanded towns sufficiently far away from the city to be self-contained. Thus at a regional level the spatial impact has been one of increased zoning and segregation of land use functions.

Because we work within a free market, planning's economic impact is difficult to judge. It's probably fair to say that planning policies have not controlled the land market as had been hoped. The main issue seems to be over the release of planning permission to develop new sites. In the south of England at least this has had mixed fortunes. Land prices have risen making profits for some but at the expense of a housing mix which would allow low-income people to buy.

Widespread development of inner city areas has led to the dispersal of communities of which London's Dockland scheme is perhaps the best example. By putting specific housing types into specific areas it has meant that the social mix of areas has reduced. We are more likely to find housing estates with homogenous social groupings. In poorer areas this has led to a feeling of a 'sink effect' with consequent social problems and unrest. Other nations, such as the USA, have long realised the negative side of **'ghettoisation'**.

The environmental impact of planning is not something often noted except in individual schemes. Large-scale development must have some form of environmental impact assessment carried out but the extent to which that determines planning policy is open to question. The zoning of land has had the impact of concentrating effects in one area which might overload environmental systems in the case of pollution. In other ways, planning has been highly positive. All the National Parks and nature reserves are a result of planning decisions. The inclusion of Heritage Coasts (outstanding coastal areas worthy of protection) has given planners the chance to control unsightly development.

1.3.6 Is planning universal?

Today, all nations have some sort of urban/rural planning. With the increasing importance of the EU, we can expect European nations' planning to become more alike: regional planning tends to be focused on jobs and job creation. Urban planning usually mixes financial incentives, regulations and public infrastructure to move land use along desired lines. However, there are some important differences and it is worth looking at a few to see what lessons we can learn.

France's main difficulty in planning is Paris. This urban area is so large and important that it has created an 'urban desert' around it. Resources have been sucked into the Paris region leaving little except problems for surrounding towns and villages (and also, of course, for Paris in trying to cope with a system grown too large). This centre-periphery problem can be seen elsewhere (e.g. Rio de Janiero) and so the way the French have tackled it could give us ideas for elsewhere. Since the 1960s, development in Paris has been checked to make sure it doesn't add to the problem. Meanwhile, other important towns in France have been given special development status with the aim of creating **growth poles** to combat Paris' hold. In some areas this has been spectacularly successful, e.g. Toulouse region with its aerospace and high-technology centres. Paris itself is subject to just one authority making city-wide planning more simple. Apart from the reconstruction work in the city there has also been the scheme to build a ring of towns around the outskirts to cope with the extra population forecast. Although the system is not perfect (Paris still receives a large share of resources) it has gone a long way to reduce the problem.

Germany has a similar problem only in this case the urban centre is not one large city but two belts of urban areas centred along the rivers Ruhr and Rhine. The urban desert of France was not seen in Germany but they did have the equivalent – a poor rural sector. Thus it would have been possible to talk of two Germanies: a prosperous urban one and a poorer rural one. By use of development assistance this problem had been largely removed by the mid-1980s. In some areas, e.g. Rheinland-Pfalz the planning authorities used the ideas of Christaller ($\Rightarrow$ settlements) as a basis for design work (only they used travel time rather than distance as a measure of central place).

Italy, like the other nations, had a spatial disparity that it chose to solve by planning means. Here the distinction is between rich North with its industry and poor South with its agriculture. The South (the Mezzogiorno) has had plans aimed at land reform and better transportation. Although this aid was limited by EU rules, Italy switched ideas in the 1960s and 1970s to use a growth pole strategy in the South. There was some success but the lack of suitable cities to act as growth poles meant that it never really took off. Furthermore, the schemes were based on public rather than private expenditure which limited community involvement and has been cited as another reason for failure.

The limited population of Scandanavia has given a different slant to planning. With most people living in a few urban centres one can afford to concentrate on local planning. For example, Copenhagen rejected the British green belt idea in favour of the green finger: growth could take place along key transport routes with the space in-between left as rural areas. With rising populations the response seems to be to add new parts to the city rather than to decentralise. Stockholm, small by European city standards, again used transport as a key planning tool only this time the underground was a key aspect. To avoid sprawl the planners kept the housing density high near the stations letting it decrease further away. Thus there are a series of zones of changing density away from major transport links. Although this has been a good example of planning there have been the inevitable problems – this time with a lack of decentralisation.

These cases have shown how different nations have reacted to the same problem – land use and urban growth. Larger populations seem to favour decentralisation whilst smaller cities can attempt to disperse growth along their transport links. Every study has demonstrated the centre-periphery problem: a rich core and a poorer surrounding area. French planners considered that this centre-periphery problem was also visible on the international scene and considered an area from Birmingham to Milan to form part of a super-centre. Italy, with its two regions shows that you can't just plan away your problems: there must be something there to start the process. Another common link is that planning is seen as a positive activity e.g. it promotes something. Only in Britain (and France in the case of Paris) is planning seen as a negative (controlling) activity.

1.3.7   Can we affect planning?

During the past twenty years there has become a realisation that planning is, or at least should be, a community activity. In this last section it is possible to see how everybody can be involved in the planning process. Many people have argued that the free market is good at developing land for profitable activities but less good at developing public functions. Thus any good plan, they say, should be a mixture of public and private money and ideas. In many developments this line is very blurred: most private planners will add some public functions such as infrastructure to a project. This activity, called **planning gain**, is the source of much argument between developers and local authority planners. To understand the modern planning system one must appreciate a number of elements involved.

Planning is a people activity; there are numerous groups involved. Firstly, there is the public planning system with the most senior authority being the Secretary of State (SoS) at the Department of the Environment. The SoS is advised by senior civil servants. Below this there may be some regional interest (although this is declining). Below that there is the local planning system. Secondly, planning doesn't just happen it requires rules and procedures to be followed. The basic route through the system can be seen in Figure 1.3.2. In addition to showing the way in which development could take place there's also mention of the different kinds of people involved at each stage (we usually refer to people doing various jobs as the actors in the development process). Thirdly, planning activities take place against a background of plans. In other words, for all major developments there is a plan which lays down what can be done where. There are two levels of plans used. **Structure Plans** are broad development outlines produced at a county council or similar level. They represent a series of statements rather than a physical plan as we would know it. Thus if it was determined that an area should be expanded then the structure plan would say which of the districts in the county should receive this growth and how much they should take. Structure plans are subject to much scrutiny. Before they can become the legal planning framework of an area they go through many consultations and modifications. The SoS finally passes the modified plans. This means that the next stage of framework planning can be designed. The **Local Plan** takes the broad outline of the Structure Plan and then puts in the detail. In our example, if the structure plan decided that a district should receive growth, the local plan would determine which areas (or streets or fields) could be developed. Again, there is a good deal of discussion but the final plans, once accepted, become the legal framework for development. Fourthly, everybody has a right to comment on a plan. The process for this varies according to what is under discussion but

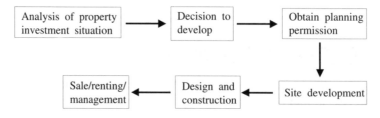

Figure 1.3.2 The property development process

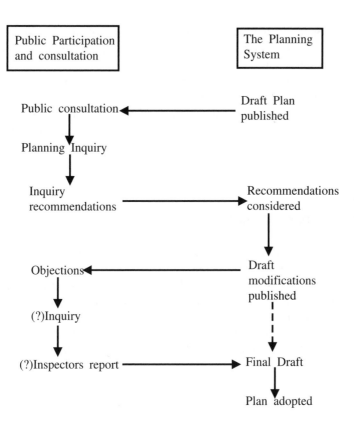

Public Participation and consultation	The Planning System

Public consultation ← Draft Plan published

↓

Planning Inquiry

↓

Inquiry recommendations → Recommendations considered

↓

Objections ← Draft modifications published

↓

(?)Inquiry

↓

(?)Inspectors report → Final Draft

↓

Plan adopted

Figure 1.3.3 *The planning process and public participation*

a look at the system for development plans (i.e. structure and local plans and similar) should give some idea of how it's done (see Figure 1.3.3). There is a repeating five-fold pattern that develops: consultation–publication–modification–inspection–adoption. Ideally, this allows for the greatest range of views to be heard whilst still keeping the system moving.

Whatever we think of the planning system it's one of the most powerful pieces of applied geography that we have today. It is important that we understand the system and are able to use it.

Questions

1. Call into your local planning department. There are usually some leaflets which explain what they do and how their department is organised. Write down the list of functions and compare this with information you get from other planning departments. What are the basic jobs carried out?

2. Gather information on the planning system in a developing (or rapidly industrialising) nation (sources could be your local library or

an embassy). What problems have they had to deal with? How have they gone about it? In what ways is this similar/dissimilar to the cases discussed here?
3. Using the local newspapers and planning department documents follow a planning issue from start to finish. How does it fit into the framework outlined in this chapter? Were there any changes? Why? What was the outcome? Would you have come to a different conclusion? Why?

1.3.8 References

Cullingworth, J. B. and Nadin, V. (1994) *Town and Country Planning in Britain*, 11th edition, Routledge.

Hall, P. (1992) *Urban and Regional Planning*, 3rd edition, Routledge.

Herington, J. (1989) *Planning Processes: An Introduction for Geographers*, Cambridge University Press.

Greed, C. (ed.). (1996) *Implementing Town Planning*, Longman.

Ward, S. V. (1994) *Planning and Urban Change*, Paul Chapman Publishing.

Subject overview: Models in geography

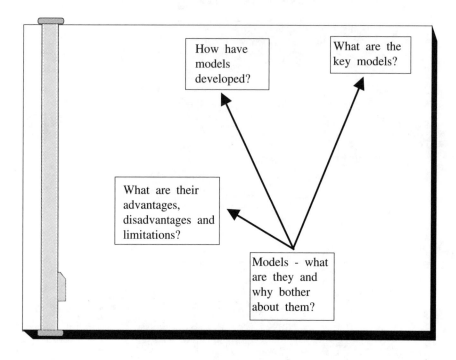

1. The value of models lies in their ability to represent reality.
2. Their main disadvantages include the difficulty of getting them to work in 'real' situations. These situations could be too complex and not deal with the right variables.
3. There are three types of model – realistic, conceptual and mathematical;
4. Realistic models, e.g. maps, are closest to our views of reality.
5. Conceptual models, e.g. flow diagrams, help us understand complex situations like ecosystems.
6. Mathematical models are used where vast amounts of data need organising.
7. All models are trade-offs with strengths and weaknesses.

1.4.1 Introduction

There are many ideas in geography which rely heavily on the concepts of models. The aim of this chapter is to examine models: to see what they can and cannot do. It will be possible to see how they can be made and used. It will also help you to be more critical of models (that is, to appreciate their advantages and disadvantages).

1.4.2 Models – what are they and why bother about them?

Models are not easy things to define. However, at the most basic level they all seem to have certain features in common:

1 They are representations of reality – they do not cover every detail, just enough to make some sense of the items under study.
2 They are abstractions of reality – their aim is to remove those parts which don't seem to affect the item under study.
3 They can be used to simplify the complex.
4 They can be used to question ideas, promote new ones or to stimulate debate.
5 Increasingly they can be used to plan actions or provide answers to theoretical questions.

Why bother with models? Couldn't we do without them? Actually, despite all the pitfalls models do have an important role to play. In any geographical situation we are going to be faced with an enormous amount of data.

Modellers try to take out the parts which appear to control the whole thing rather than allow every aspect to be included. From this we can get the broad principles that seem to govern the subject.

1.4.3 What are their advantages, disadvantages and limitations?

Too often models are presented in geography without regard to their advantages, disadvantages and limitations. This means that models are not subjected to the same critical thought that one might give to a table of data for example. Table 1.4.1 outlines main advantages and disadvantages.

There is, of course, no reason to assume that the situation is suitable for modelling. There is the assumption in Western-oriented science that there are solutions 'out there' waiting for someone to discover them. Now we are starting to look at areas which do not behave in set ways (chaos theory is one example here) we can begin to appreciate that there may be times when a model is not appropriate. Climatologists are one group of workers who see that old-style models are not useful and they are replacing them with series of equations which brings up the question of the limitations of models:

- The situations might be too complex to model and the boundaries too difficult to define.
- The workings of the area might not be understood. There must be some basic ideas before any model can be constructed.
- Models might have built-in problems.

Table 1.4.1 Advantages and disadvantages of models

Advantages	Disadvantages
The ability to tackle complex situations by producing a simplified version	Models are difficult to 'operationalise' which means they look great on paper but might no get further because the data cannot be found or measured
Producing ideas that work at both global and local scales	The most complex are very difficult to test
Can be used in different areas of geography to bring coherence	Data may not be available or data are subjective (based on opinion not experiment)
Reduce fieldwork costs by planning ahead	It may not be possible to say precisely where the model ends
It's possible to produce a list of factors which could be investigated separately	The model might be so broad that it is impossible to construct a reasonable version

- The model might not be the right one for the situation. It requires a good deal of thought before the correct model can be applied.
- The assumptions needed in every model (usually to reduce the number of items the model needs to cope with) may or may not be appropriate.

Tip

Be critical! Every model has its pros and cons – the best way to start thinking critically about geography is to appreciate this.

1.4.4   How have models developed?

The aim here is to present a simple classification so that it is possible to see how the same ideas have been applied to diverse areas of geography. Despite the vast number of models each can be put into one of three categories. The first category is the **realistic models** which would include photographs, drawings and maps. Next comes the **conceptual models**. These models are abstractions where only the main items thought to be important are left in. Examples range from a description such as a series of directions to energy flow diagrams showing only the key features of the system. The vast majority of models falls into this group. Lastly, there are the **mathematical models**. This model group contains the most abstract ideas where basic interactions are broken down into a series of mathematical formulae.

1.4.5   What are the key models?

In this section only a few of the most important examples will be described to illustrate the categories noted above.

One example of the realistic model is the procedure associated with environmental impact assessment. The modern process started around 1969 when the USA brought in a National Environmental Policy Act. This Act produced a simple model – a procedure that had to be followed. Despite criticisms this type of model spread quickly with later examples extending the complexity of the ideas.

Figure 1.4.1 shows a flow diagram for impact assessment – a more complex example. Here there are a series of goals which should be reached in order to evaluate fully the project and its effects.

The systems diagram must be one of the most abused and used of the conceptual models. It has been used in all sorts of situations, not always appropriate for it. Figure 1.4.2 is a simple input-output system ($\Rightarrow$ agri-

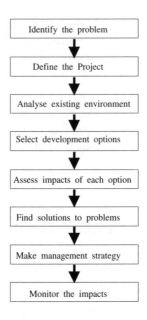

Figure 1.4.1   Environmental impact assessment model

culture to see more detailed examples). Note the key elements: input (what goes into a system to alter it), output (the result of the input's actions on the system), feedback (how the output can affect the input), store (where material for example can be temporarily taken out of usage), a sink (where it is stored long term) and a regulator (something that alters the flow). The system itself (e.g. an ecosystem) is represented as black box (where the detailed workings are unknown) although it is possible to break this down into a series of subsystems.

The final example comes from the range of mathematical models. One of the better known ones is the S-curve: a model demonstrating how a

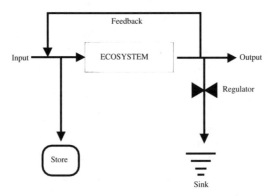

Figure 1.4.2   Simple input-output systems model

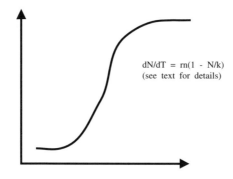

$$dN/dT = rn(1 - N/k)$$
(see text for details)

Figure 1.4.3   *S-shaped curve from a mathematical model*

population can grow. This can be seen in either equation form or as a graph
(Figure 1.4.3). In the equation N = population size, k is the maximum
population size possible, r is the growth rate and T = time. The graph
shows what happens if this equation is plotted: steep increase followed by
a smaller increase and finally a levelling off.

Not all models work perfectly. One of the best examples of this is the
'Limits to Growth' model published in 1972 as part of the debate for the
first UN Environmental Conference in Stockholm. Its basic idea was that
if people continued to use the planet in the same way as they had, then we
would be headed for disaster (see Figure 1.4.4). The idea was to create
something that could be continuously altered until it came as close to
reality as possible. Then the program could be run to see what would
happen in the future. It was argued that by using a computer model the

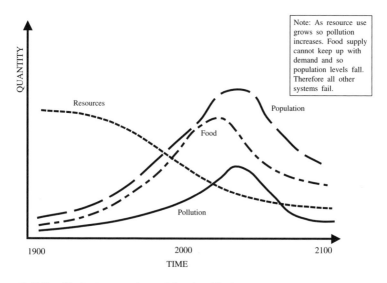

Note: As resource use grows so pollution increases. Food supply cannot keep up with demand and so population levels fall. Therefore all other systems fail.

Figure 1.4.4   *Limits to growth model – simplified*

subjective part of analysis could be removed. Once it was published there was much debate both for and against this model. One example of this was to note that when the model was run backwards in time it did not mirror events: therefore it must be a faulty design. In fact, it wasn't designed to go backwards and so no wonder it didn't work! Other arguments examined the assumptions underlying the model, the mathematical formulae and the very complexity of the model (making it virtually impossible to check). The main point here is that any model has its pluses and minuses – the aim is to find what they are and to be able to argue critically.

1.4.6   References

Gilpin, A. (1995) *Environmental Impact Assessment*, Cambridge University Press.

Huggett, R. J. (1993) *Modelling the Human Impact on Nature*, Oxford University Press.

Jakeman, A. J., Beck, M. B. and McAleer, M. J. (1995) *Modelling Change in Environmental Systems*, Wiley and Sons.

Smith, L. G. (1993) *Impact Assessment and Sustainable Resource Management*, Longman.

1.5 History and philosophy of geography

Subject overview: History and philosophy of geography

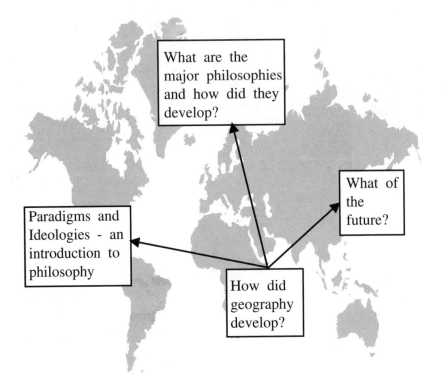

What are the major philosophies and how did they develop?

What of the future?

Paradigms and Ideologies - an introduction to philosophy

How did geography develop?

1. The history of geography is a complex interweaving of four factors: place, people, time and inter-relationships.
2. Earliest geographers used geography for studying people, places, astronomy and religion.
3. The end of the Roman Empire stopped the spread of geography with Medieval monks taking some important documents.
4. The biggest boost for geography was fifteenth-century exploration with key figures, e.g. Columbus, becoming increasingly important.
5. Although exploration died down by the eighteenth century other aspects became important such as landscapes.
6. The nineteenth century saw geography develop as a university and specialist subject.
7. In the twentieth century geography has split into two main divisions – human and physical – and two main approaches – positivistic (science/models) and behavioural. Currently, the subject is expanding rapidly the approaches it uses.
8. The value of philosophy is that it should make us question everything and not accept ideas blindly. The aim of philosophy is to allow us to explore ideas through logical enquiry and critical thinking.
9. Major groups of ideas can be collected under a paradigm – a way of thinking and working, followed by most workers in the subject. Groups of opinions, attitudes and theories that follow a common set of rules makes up an ideology.
10. Individual philosophies have three aspects: epistemology (what is known), ontology (how we can know it) and methodology (how we can test it). There are three basic philosophical groups: empiricism/positivism, idealism/humanism and structuralism.

1.5.1 Introduction

The aim of this chapter is to provide the reader with a brief summary of key developments in the history and philosophy of the subject. There are three benefits in doing this:

1 The ability to see how an idea has developed through time. Urban modelling is a good example (see Chapter 2.2). Models which appear disconnected fall into an understandable sequence of ideas.
2 The chance to study the growth of the subject and appreciate the way in which its ideas have developed.
3 The opportunity to link geographical ideas to people and places. Like any subject, geography has its key people and key thoughts. It enriches

the study of the subject to see who these people are and what they did.

Geography is not a linear subject, it keeps reworking past ideas. Developments are often linked with a particular place, or set of people or even a specific time. By looking at these aspects it is possible to follow the development of an idea (which aids understanding). From a study of place it is possible to find key research centres. A study of people brings out the key workers in the field (try checking a few bibliographies on a topic and soon the same people will be cited – check out these people). The time factor is involved in tracing work. Is this the first time an idea has been mentioned or is it just reworking old themes?

In a subject as diverse as geography, philosophy, the study of knowledge and ideas, can provide the only link between widely differing parts. Furthermore it can help us understand the way the subject has developed and the theoretical basis on which it was founded. Philosophy:

- Provides a framework – much like a set of rules and procedures. Because some ideas, especially in human geography, are questions of interpretation, then how we interpret it (i.e. what philosophy we use) is vitally important.
- Forces us to think about what we are doing to avoid bias in our work.
- Helps us to understand the idea of true/false statements. Much of geography is taken up with human understanding about the environment. This has meant that there's rarely been any agreement about what can be accepted as true. For example, does the 'region' exist or is it just a convenient term to group things together?
- Helps make geography a more critical exercise. Basically, it means that argue everything through and accept nothing without a thorough check. Sadly, much of what's written (in either books or journals) is too uncritical. It gathers ideas but does not put them into a strict framework. Since many writers argue that great leaps forward in knowledge are made by small well-argued steps then it makes sense to use the framework we've been given.

1.5.2 How did geography develop?

Although geography as an academic subject recognised by universities is only about 100 years old, the use of the subject stretches back for thousands of years. In this section we look at the very earliest period – classical Greece and Rome primarily with Arabian and Chinese contributions noted. Figure 1.5.1 is an attempt to map this diverse topic. The lines linking key books/people/ideas are there to suggest the ways in which our understanding of geography has changed and where the changes have come from.

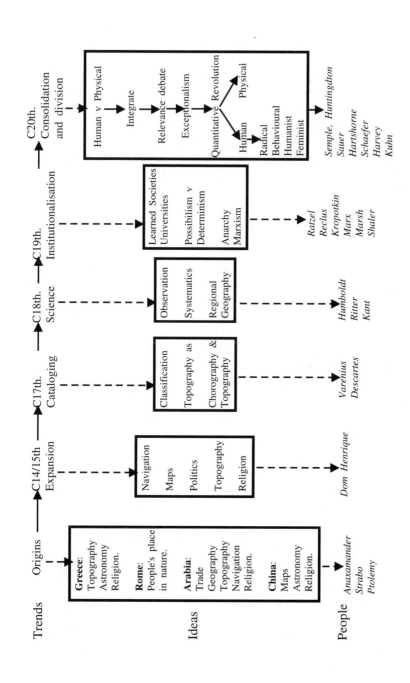

Figure 1.5.1 Timeline of the history of geography

Much of the early work by the Greeks was concentrated on topography (people and places), astronomy (the workings of the heavens) and religion. The Roman Empire also considered geography to be important: much work centred upon the study of people's place in nature. Surviving works from the first 'true' geographer, Strabo (60BC – AD21?) show a considerable series of descriptions from what we would call economic and social geography to biogeography. A completely different form of geography was provided by Ptolemy. His first work, an astronomical guide called the *Almagest* was followed by *Geography* – a series of latitudes and longitudes for a vast number of regions and features. So comprehensive was this mathematical treatment that it was used hundreds of years later to produce a series of maps which fuelled fourteenth-century maritime exploration. Arab and Chinese influence was less but it is known that both groups needed accurate maps suggesting a strong geographical background. By the thirteenth century the decline in geography noted from the fall of the Roman Empire was reversed because of the need for maps in exploration. New states were forming of which Portugal, Spain and Venice were the three most powerful. Each accepted that the best way to gain power was through trade. This highly practical use of geography continued until the seventeenth century when people such as Kant (1724–1804) tried to refine the subject through the application of new philosophical ideas. These developments were extended in the eighteenth century by people such as Alexander von Humboldt (1769–1859) and the more theoretical Carl Ritter (1779–1859). Where von Humboldt concentrated on systematics and physical regional geography, Ritter was focused by his strong religious convictions into a human-centred regional geography.

Ritter was also concerned about education. He combined these, through his post as the only German professor of geography at the time, to produce a geographical education for school pupils which emphasised the local area. Throughout the nineteenth century both university geography posts (e.g. Oxford in 1887) and learned societies (e.g. Royal Geographical Society in 1830) were formed. During the nineteenth century it was far more popular as a school subject where 'commercial geography' – the study of the location and resources of Empire – helped to give civil servants some idea of their new postings!

The divisions between physical and human geography seen in the nineteenth century were to continue. By the 1920s this split was seen as more than just a minor one: geography was to mean human ecology with any physical aspect being taken as geomorphology. Some geographers (e.g. Carl Sauer) argued for integration whilst other geographers (e.g. France's Vidal de la Blache) were concerned about links with other subjects like sociology. As geography developed so did its sophistication. Behind all this there was the main question: 'what **is** geography?' The problem was that there had been little debate about the theoretical nature of the subject. This changed with the publication in 1939 of Richard Hartshorne's 'The Nature of Geography: a critical survey of current thought in the light of the past'. To Hartshorne the answer was that geography was a holistic study – taking something from all of the sciences (because landscapes, the 'laboratories' of geography, were complex places where individual items could not be separated from the whole). The Second World War put an end to the debates for a while as geography regained some of its former importance by becoming a vital part of the war machinery (in much the same way that the map-makers had in the fifteenth century).

The 1950s saw a resurgence of the debate. This time the division was between integrated (regional) studies and a more 'scientific' systematic approach. Slowly the systematic approach became more popular. By the start of the 1960s the 'quantitative revolution' (a method which assumed that all geographical phenomena could be explained by the use of mathematical and systems models) became popular. Seeking more theoretical ideas, US and UK geographers started to use continental European work (something impossible during the 1930s/1940s). By 1967 these ideas had gained sufficient ground to be considered the new way of looking at geography. Two British geographers, Richard Chorley and Peter Haggett edited one of the most influential texts for this new scientific geography *Models in Geography*: the title showing how the new work was to be seen – a theory-based study of people and place.

No sooner had some of the groundwork for this quantitative revolution been laid than it was under attack. The principal complaint was that this new scientific geography added nothing to our understanding of the subject. Geographers were increasingly demanding what their work could do to bring about change. Labelled the 'relevance debate' this should be seen as just one expression of a growing concern. The reaction to these words was not to remodel quantitative geography but to move into new directions with a far greater emphasis on theory. This new 'behavioural geography' argued that people's perceptions were as important as reality. Human geography moved into more radical territory (leaving physical geography with the scientific orientation to its work). The whole idea of radical geography was to create a subject that didn't just describe, it would help change people's views about themselves and the world they lived in. The 1980s saw a growth of geographical writings examining these, and similar ideas. Recently, a new philosophical idea, postmodernism, has been seen. Although it's too early to say how postmodernism might develop there is some agreement that it is about the differences in space i.e. trying to get away from the idea of 'grand theory' seen in both scientific and humanist geographies.

1.5.3 Paradigms and ideologies – an introduction to philosophy

An American physicist, Thomas Kuhn, had been studying how subjects changed. He argued that each subject has a set of basic rules (which he called a **paradigm**) that everybody in the subject follows. Imagine that over time a series of discoveries makes the existing paradigm less successful as a way of working. Then, according to Kuhn, there will be a 'revolution' and a new paradigm will be constructed. Such a **paradigm shift** as it is called has happened several times to geography in the last 50 years. One of the most important shifts was in the introduction of systems models into human geography. This 'quantitative revolution' as it became called was fundamental because it shifted geography away from a paradigm of descriptions and towards a paradigm that people were capable of being seen as acting like mechanisms. Some time afterwards, another shift changed human geography towards a more idealistic philosophy. The value of the paradigm can be seen in its ability to help us explain why certain 'schools' of geography exist and how they operate as they do.

If the paradigm defines the kind of prevailing thoughts about a subject then **ideology** puts it in a wider context. Paradigms are about the place of knowledge in geography; ideology is about the place of geography in society. Ideology is about the opinions, attitudes and theories developed by one group of people to defend their interests. For example, take British politics. There are two main groups – Conservative, based on capitalism and Labour, based on socialism. These two ideas, capitalism and socialism, would be considered as competing ideologies. Today, divisions in ideology can be seen clearly in geography and have helped to develop the ideas of the subject. Current examples include feminist and Black geographies – each has their own perspective of looking at the world.

1.5.4 What are the major philosophies and what do they do?

In philosophy as in geography there is not one right answer but several. Thus there are several different philosophies each with a common framework but each trying to describe the world according to its own rules and procedures. Think of this as a matrix (see Table 1.5.1). Each philosophy has its own **epistemology**, **ontology** and **methodology**. Epistemology is the theory of knowledge. It answers such basic questions as what can we know and how we can know it. For example, if we say that we only accept things we can sense then if we can sense something, it must exist. Related to epistemology, ontology is the theory of existence or what can be known and what can exist. Therefore if something exists we will experience it. The easiest of the three parts, methodology, is the ground rules and procedures to test for existence.

Although there are a large number of different philosophies, in practice they can be placed into one of three main groups. Within these groups there is a basic agreement as to how the philosophy works but each division will emphasise its own particular idea. The main groups are:

- **empiricist/positivist** These philosophies work on the notion that we know things through experience (epistemology) and that the only things we experience are the things that exist (ontology). This leaves us with a methodology that is concerned with the production of verifiable evidence, e.g. experimentation. Philosophies in this group emphasise the scientific approach to working. This group includes empiricism, pragmatism and positivism.
- **idealist/humanist** These philosophies argue that everything is based on perception – reality is a mental construction. Thus the epistemology argues that knowledge is created from subjectivity whilst the ontology states that what is perceived to exist, exists. The methodology would be examining subjective ideas and individuals rather than the theory-led, model-building approach of the empiricists. This group includes idealism, phenomenology and existentialism.
- **structuralist** This group is characterised by sharing an ontology which argues that what exists cannot be observed directly. A corresponding epistemology would note that appearances might not be true and that observations would not reveal the whole picture. Perhaps the best way of looking at these philosophies is to consider the puppet. What we see is a small wooden doll moving about. In reality, the puppet is connected to strings from which the puppeteer controls all the actions. Structuralist approaches would argue that the 'truth' is not the puppet but the puppeteer. This group includes structuralism, realism and functionalism.

1.5.5 What of the future?

How will all this turn out? For the last twenty years, geography has gone from a move to create a complete scientific subject to one where philosophy, diversity and social concern are seen as the key ideas. Potential trends for the future include:

1 The divisions between physical and human geography seem to be complete without either side being particularly concerned. Physical geography becomes a science whilst human geography moves firmly into the social science realm.
2 Human geography seems to be continuing along the lines of a social-concern subject. This implies a move towards greater philosophical

Table 1.5.1 Major philosophies and their characteristics

Philosophy	Epistemology (E), Ontology (O) and Methodology (M)	Principles	Critique	Major writers
Empiricism	(E)Knowledge gained through experience; (O)We experience what exists, (M)To present experienced facts	Only observational statements are valid. Emphasises the role of scientific inquiry	Data is allowed to define problems. Evidence is not submitted to scrutiny. Rejects theory in favour of observation	Bacon von Humboldt Ritter Locke
Pragmatism	(E)Concrete experience; (M) hypothetico-deductive	Defines meaning and knowledge in terms of function of experience	Emphasises the human element. Concentrates on practical aspects	Dewey James
Positivism	(E)Knowledge gained through experience which must be agreed on; (M)Verification of factual statements	Theory-led. value-free. Progressive unification of laws into a single system. Scientific method central (especially logical positivism)	Links geography with other subjects. Provides a rigorous base to study. Rigid adherence to science	Harvey Comte von Thunen Sauer
Idealism	(E)Knowledge is subjective; (O)What exists is that which is perceived; (M) Reconstruction of contexts	Knowledge does not exist without the knower. Reality is a mental construction. Centrality of beflef	Studies rational actions of humans. Permits study of motives. People not dehumanised. Separates effect from causes	Guelke Dilthey Vidal de la Blache Kirk
Phenomen-ology	(M)To analyse and identify basic features of subjective	No independent world outside human existence. Study free of presuppositions	Based on emotion but impossible to enter the mind of the subject. Focuses on experience	Husserl Hegel Buttimer

Table 1.5.1 Continued

Philosophy	Epistemology (E), Ontology (O) and Methodology (M)	Principles	Critique	Major writers
Existentialism	(O)Efforts to overcome detachments; individualistic	Reality created by free acts of human agents. Does not believe in ultimatete knowledge (opposite of phenomenology)	Seeks to bridge gap between subjective and objective. Enables geographers to explore shared meaning of landscapes	Satre Heidigger Nietzsche Samuels
Structuralism	(E)Mechanisins not revealed to the world., (O)What exists cannot be observed directly; (M) Construction of theories	Theoretical articulation to find universal structures which provide motive forces in society. Mix of observation and theory	Concern for socio-economic reality. Can expose the entirety of a system. Has been used a the basis for political systems (Marxism)	Levi-Strauss Marx Althusser Habermas
Realism	(E)Nothing exists except by objective senses; (O)Nothing exists which cannot be observed; (M)Study independent of people	Mechanisms are independent from the events they generate. Phenomena can be explained by recourse to structures	Attempts to look at causal mechanisms. Argues against positivism and idealism and their limitations	Bliaskar Keat Urry Nunn
Functionalism	(E)What we see is constructed of functions we cannot see directly; (O)Functions can be seen but not explained; (M)Fieldwork the only way to collect data	Concerned with flinctions and their analysis. Stresses systematic properties of groups. Society is holistic	Used as a research tool in many social sciences. No agreed standard definition. Static system – cannot account for change	Eisenstadt Darwin Vidal de la Blache Durkheim Parsons

debate whilst at the same time broadening the 'acceptable' number of approaches, e.g. feminism, post-modernism.

3 If concerns about environmental degradation continue then it seems likely that geography will take up some of the research. There are already pressures for geographic science to provide more than models when faced with pollution, urban development etc.

4 All these moves will take place in conditions of increasing financial difficulties. The increase in the number of students allied with a decrease in funding has meant that many universities are being very cautious about spending. This means that if research is to be carried out it will need to be funded by outside agencies rather than the university. In turn, this implies that some more radical ideas may fail to find funding.

1.5.8   References

Gregory, D., Martin, R. and Smith, G. (eds.). (1994) *Human Geography: Society, Space and Social Science*, Macmillan.

Gregory, D. (1978) *Ideology, Science and Human Geography*, Hutchinson.

Haggett, P. (1990) *The Geographer's Art*, Basil Blackwell.

Johnston, R. (1983) *Geography and Geographers*, 2nd edition. Arnold.

Johnston, R. J. (1986) *Philosophy and Human Geography*, 2nd edition, Arnold.

Rogers, A., Viles, H. and Goudie, A. (eds.) (1992) *The Student's Companion to Geography*, Blackwell.

Unwin, T. (1992) *The Place of Geography*, Longman.

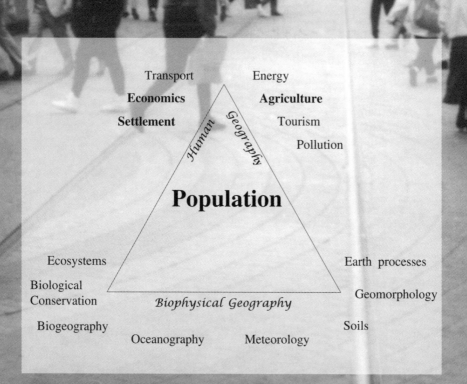

Transport   Energy

Economics   **Agriculture**

Settlement   *Human* *Geography*   Tourism

Pollution

Population

Ecosystems   Earth processes

Biological
Conservation   Geomorphology

Biogeography   Soils

Biophysical Geography

Oceanography   Meteorology

Subject overview: population

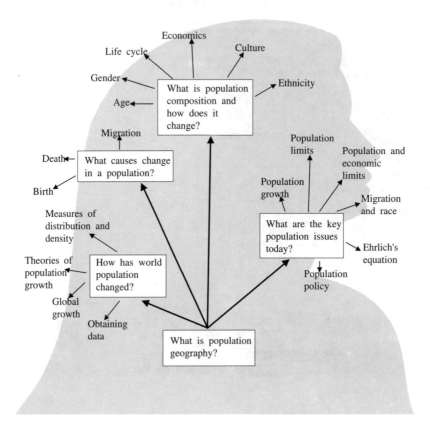

1. Human population, its size, distribution and impact is one of the most important issues in modern geography.
3. World population increase probably started in a number of centres. Its growth has been most pronounced in the twentieth century, where developing nations are providing greatest increases.
3. Governments need to monitor population levels for a number of reasons – financial, social and planning. Techniques available range from the census and registration to indirect evidence. Cost is a major factor.
4. Population impact is not uniform but increases where population density is greatest.
5. Current theories accept the demographic transition model as a model of population growth linking economics, health and nutrition to changes in birth and death rates. Earlier models such as Malthus, emphasised food production.
6. Population numbers change according to:
$$P_1 = P_0 + B + I - D - E$$
7. Population characteristics are usually based on age, gender, birth rate, fertility and death rate.
8. Migration, for whatever reason, can alter dramatically population levels (especially with refugees). Migration models usually emphasise the cost-benefit approach.
9. Population issues are generally controversial. Current ones include population growth/capacity, race/segregation and impact on the environment.

2.1.1   Introduction

Population geography is concerned about the ways in which the human population changes through time and space – in other words, distribution and density. The answers to these questions are of vital importance as we try to fit an ever-increasing human population into an area which is not just the same but actually declining as the demands we put on it lead to its degradation.

2.1.2   What is population geography?

In simple terms, population geography is the study of the distribution of people. In practice, this involves a range of specialist subjects such as

demography (the statistical analysis of population) and **population modelling** (how and why populations change as they do). This makes population geography one of the most interesting and important studies. Not only are we concerned about numbers and processes but also about the implications of these and how we view them.

2.1.3　How has the world's population changed?

One of the most common diagrams that any geographer comes across is the exponential curve representing population growth through time which, however you put the scale, rockets upwards just at the turn of the century. Although this is quite an accurate graph (as far as we can tell for all our limited data) it really doesn't give the whole picture. The aim of this section is to present a view of the world's population changes which challenges this simple idea. There are three aspects we need to study: global growth, theories of population growth, and measures of distribution and density.

These aspects depend on data. Firstly we can use a census. This is a form sent out (in the UK) every ten years which aims to get a range of information from everybody. Questions go from age/gender to employment and travel patterns. Censuses can produce a great deal of information but they're expensive (some nations can't afford it) and some people can get left out. Secondly, there's registration – the demand that births, deaths and marriages are recorded by the state. This gives us a continuous flow of information but only about a very limited range of data. Thirdly, one could use sampling. Alternatively, for populations subject to movement, migration statistics can be kept. Finally, for past populations there's the indirect evidence left behind in records and on archaeological sites.

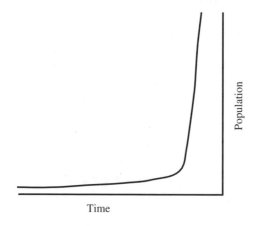

Figure 2.1.1　Exponential growth for population

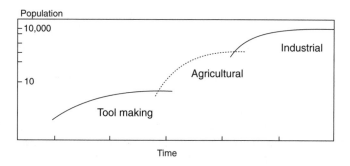

Figure 2.1.2 Logarithmic scale for population growth

Such information is all we have for most of human existence and its interpretation can be difficult.

Although there is much talk of a 'global' population explosion, the picture is not that simple. For example, rather than have a single centre for population growth it is thought that there were several (e.g. Africa, Middle East) mirroring the development of agriculture. Furthermore, detailed studies in Saudi Arabia, Mexico and elsewhere point to regional population levels which expand and contract in response to environmental and social conditions implying global growth can be seen as an aggregate of a series of regional growth/decline patterns. Each region varies but over the centuries the total trend is upwards. Is this an even trend? The answer depends on how the data are presented. Arithmetic presentation of the data presents the usual exponential curve (Figure 2.1.1) but using a logarithmic scale produces a series of rises and plateaux (Figure 2.1.2). Demographers argue that this corresponds to an increase in resources which would in turn support a bigger population. The three 'events' which would give this pattern could be seen as the 'Stone Age', 'Agrarian' and 'Industrial' revolutions. The interesting thing about this model is that if you

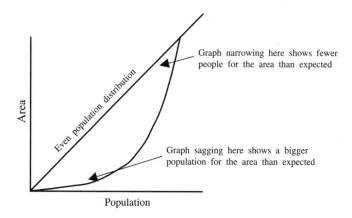

Figure 2.1.3 The Lorenz curve

accept it then you assume that the current rise in population will level off of its own accord. These two graphical techniques look at change through time. It can also be shown that population varies with space – some areas have more/less population than one would expect. A cumulative graph of population and area (a **Lorenz curve** – Figure 2.1.3) highlights population 'light'/'heavy' areas.

The next element is to consider some of the mechanisms that might provide an explanation for the patterns we've studied. One model, the **Demographic Transition Theory** (DTT) offers the best explanation. The DTT is based on the study of changes in birth and death rates for a given area through time (Figure 2.1.4). The graph is divided into four stages each one of which represents certain population characteristics. Stage 1 is a time of high birth and death rates due mainly to low life expectancy and poor health care. Although these two figures are high they tend to cancel one another out and population remains low. Stage 2 is seen as the introduction of health care and better nutrition which lowers the death rate but leaves the birth rate high. The population level will rise quite quickly. In stage 3 there is a change of birth rate due to an improvement in economic conditions leading to people having fewer children. The population still rises but less quickly. By stage 4 the birth rate and death rates are roughly equal and the population stabilises.

The DTT might be the current explanation but there are others. In 1798 the Rev. Thomas **Malthus** published his *Essay on the Principle of Population*. He argued that while population would increase geometrically agricultural production would only increase arithmetically. At some stage the population would outgrow the food supply and 'positive checks' such as famine and disease would work to reduce population levels. This would not be immediate; there would be some 'overshoot' followed by variation around a stable figure (see Figure 2.1.5).

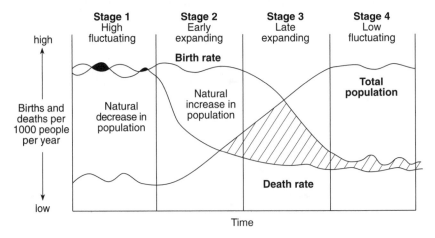

Figure 2.1.4 The demographic transition theory model

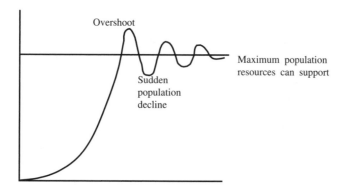

Figure 2.1.5   Population overshoot

The final aspect to be considered is population distribution (where are they) and density (how many in one place). To express the idea of distribution and density we can use ways other than maps. One of the more common ones is the **hypsometric curve** – a graph showing the distribution of population vertically (from sea level). Figure 2.1.6 shows two such graphs: one for the percentage at a given height, the other for the density. Both show the one key trend – most of the world's population lives within 500 metres of sea level. Maps also show that most live within 100 km of the coast.

2.1.4   What causes change within a population?

At the heart of this is the basic equation of change:

$$P_1 = P_0 + B + I - D - E$$

which means the population (P_1) you get at the end of year depends upon what it was at the beginning (P_0) plus births (B) and immigrants (I) (arrivals) less deaths (D) and emigrants (E) (departures). The equation identifies three areas of population change: birth, death and migration. For each of these areas data are needed alongside an explanation of variation. For example, not only is it crucial to know about the numbers of births but it is important to know why they are changing.

As with overall studies of population, data are essential. For birth rate, this usually means:

1 Crude birth rate (CBR) – the simplest of all the measures being the number of births (B) or deaths (D) per thousand of the population (P): CBR = B/P × 1000.

2 Age specific birth rate (ASBR) – a useful measure because it compares

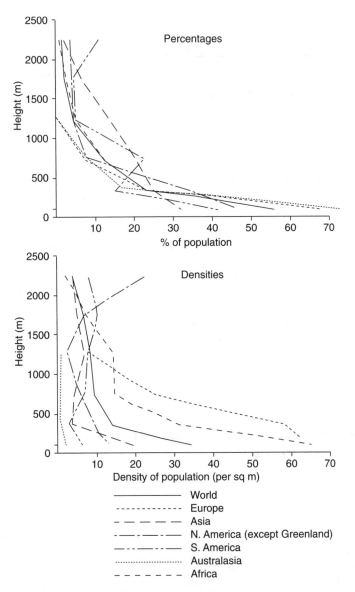

Figure 2.1.6 Vertical distribution of population

the number of children that women at a given age are having (Ba) with the number of women at that age (Pa): ASBR = Ba/Pa × 1000.

3 General fertility rate (GFR) – deals only with women who could have children i.e. 15–44 years old: GFR = B/P[15–44] × 1000

4 Total fertility rate (TFR) – an idealised figure that assumes that the proportion of women that might have children (calculated from the ASBR) actually do have them. The value of this is that it's possible to

see if actual births are below or above this figure which would suggest whether the population is increasing, stable or actually declining.

All of these measures are fairly simple: it is possible to make the matter more complex. For example, recruitment ratio could be used (a measure more common in ecology). This can be defined as the number of women who survive to reach reproductive age. This is useful when comparing developed and developing nations where fertility might well be affected by nutrition for example.

Having gained the data there is need for explanation to the question: why does the birth rate vary? The vast majority of births still occur between couples who are joined through legal and/or religious ceremonies. It follows that anything that changes that status might be a cause of variation, e.g. age of marriage (which alters the number of reproductive years), marriage-inheritance theory (where people won't marry unless there's enough money). In the **economic theory of fertility** women won't have children if they feel they might loose more by loss of income than they might gain from having children. Thirdly, there are links to industrialisation and urbanisation. The movement of people into towns and the growth of the domestic economy have both been linked to increases in the birth rate although there's the counter argument that increasing wealth leads to birth rate decline (see above). Fourthly, there is some evidence to support a difference in birth rate between urban and rural people. Possible explanations include the greater amount of land in rural areas (so children can have enough to farm) or poorer economic conditions (making children a useful source of security and income). Fifthly, health and nutrition conditions are critical for the proper development of the baby. The better the birth conditions the greater the chance of survival. It should not be forgotten that in many societies (but not all) the woman does have some choice as to whether or not to have children. Part of this choice would involve the presence of contraception. This has been one of the great factors in population decline in much of the world. It is also highly controversial. Some religions ban contraception. In some nations desperate to shrink their populations there have been stories of forced sterilisation of men. Finally, social class could be a variant in birth rate with higher socio-economc classes giving birth to fewer children than lower classes.

The study of death rates follows a similar pattern. Crude death rate and age-specific death rate can be calculated as: $CDR = D/P \times 1000$ (where D = Number of deaths) and $ASDR = Da/Pa \times 1000$ but you can also have age and sex specific death rate where $AaSDR = Das/Pas \times 1000$. In addition, there is the standardised death rate (also calculated in the same fashion as ASBR). More importantly there are two other measures which are very useful. The infant mortality rate is a measure of the number of infants (usually taken as 0–4 years old but can be just born (0 years or neonatal age), and is: $IMR = D/B \times 1000$. Lastly, there is the expectation of life at birth – 'life tables'. Used by insurance companies in trying to determine how long, on average, we are likely to live (and therefore how much insurance premium we need to pay) these can show interesting variations especially between class and nation.

Turning to those factors which cause the death rate to vary one finds similarity with birth rate factors: factors which make for a successful birth will also ensure a longer life! One of the key factors is better food/nutrition. The key to a longer life is not just a good supply of food but also enough of the right kinds of food. From the eighteenth century onwards in the UK and in many parts of Europe, the quality and quantity of food increased. This meant that people were generally fitter and able to live longer. Secondly there is increase in manufacturing. It was argued that an increase in manufactured goods means more clothes, better sanitation. In other words, part of the profits of the Industrial Revolution went, in effect, to increasing the lifespan of the population. This is linked to better health conditions. During the nineteenth century the linkages between poor water quality and disease were being found. Vaccinations became increasingly common and one, for smallpox, had a tremendous effect on lowering the death rate from this disease. In a related action, public health was developed.

The factors mentioned above have all played their part in lowering the death rate and, therefore, allowing the population to expand in accordance with the ideas of the DTT. One aspect less often mentioned in these contexts is the cause of death. Several studies have shown that the causes of death vary with age, gender, lifestyle, season etc. Furthermore, these causes seem to vary with area not just between nations but within them. This has opened up a whole new area of study called epidemiology. Apart from the changes in cause of death with age (e.g. young males are more likely to die from accident or violence in the UK than any other group) there is some suggestion that the causes of death vary with the level of development of a nation. Compare the common causes of death in developing and developed nations with the **Epidemiological Transition Theory** (ETT) – see Figure 2.1.7. In a society with very poor health facilities the most likely cause of death will be famine or plague. By the time you get to richer nations with better general healthcare the main cause of death is cancer.

Finally, we come to the third of the factors which causes population numbers to vary – migration. One of the principal problems is in deciding just what counts as migration. Researchers in this field have identified a number of variables which need to be taken into account in producing a good analysis of migration. These include distance, time, boundaries, area, external influence, reasons, numbers and social class. Because migration has so many variables, many workers have tried to make simple models which allow as many of these variables to be taken into account as possible. The major problem with all these is that, unlike birth and death, migration is a highly individual thing. Two identical people may make two opposite choices (e.g. stay or go) and both be seen as equally valid. What

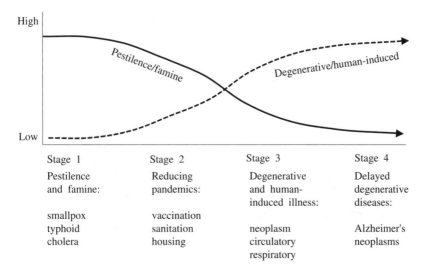

Stage 1	Stage 2	Stage 3	Stage 4
Pestilence and famine:	Reducing pandemics:	Degenerative and human-induced illness:	Delayed degenerative diseases:
smallpox typhoid cholera	vaccination sanitation housing	neoplasm circulatory respiratory	Alzheimer's neoplasms

Figure 2.1.7 *The epidemiological transition model*

most models try to do therefore is to group together millions of individual decisions and see what broad-scale patterns emerge. Six basic migration models can be described.

1 **Zelinsky's Mobility Transition**. This can be linked to the demographic transition theory. As a nation goes through development so the nature of migration changes (Figure 2.1.8). In early societies there's very little migration even to the next village. In stage 2 there's a movement from rural to urban areas and some expansion into new areas if that's possible. By the time you've reached stage 5 there will only be some urban-urban migration and a restriction on immigration through migration policies.

2 **Ravenstein's Laws of Migration**. These are some of the most widely quoted ideas on migration. They are based on the attractions of urban areas. His basic ideas included:
   i)   most people only go a short distance
   ii)   migration goes step by step
   iii) long-distance migration will be to large industrial centres
   iv)   emigration will have a compensating immigration
   v)   migrants tend to be urban adults
   vi)   migration increases with population size
   vii) large towns grow by migration rather than natural increase and people tend to go from rural to urban areas.

Although these ideas came from work in England in the nineteenth century many of his ideas are still valid.

3 The **intervening opportunities** models. Many people have followed the work of Ravenstein. Two such workers have been Lee and Stouffer.

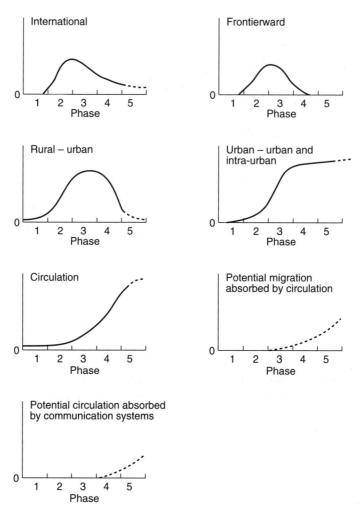

Figure 2.1.8 Zelinsky's mobility transition

They both made similar models which together can be seen as 'intervening opportunities'. Basically, this model argues that people will migrate from A to B if there is sufficient reason and unless there are better opportunities in between. This means that there are four sets of factor to take into account: A's attractions, B's attractions, intervening attractions and personal preferences.

4 **Distance-decay models**. The number of migrants decreases the further you go from the town. The resulting distance-decay curve is one of the key ideas in geography and can be used in other studies.

5 **Gravity models**. Such models look at the size of the towns and use the universal gravitation equation to calculate the number of migrants.

Thus: $M = Pa \times Pb/D$ where M = number of migrations, Pa and Pb are the populations of towns, a and b and D the distance between them.

6 **Push-pull factors of migration**. Given the complexity of the decision-making process in migration some people have preferred to use a very simple 'balance sheet' to examine migration patterns. Basically, you put on one side all the factors pushing you out of your present location and those pulling you into it and on the other side, those factors pulling you to a new location and those pushing you away from it. According to general theory, you'll migrate if the benefits of moving outweigh the disbenefits.

Finally, a look at some of the causes of migration. Some of these will fit into models, others are of a more practical nature. Given that the homelessness from migration is one of the most pressing global population problems it is crucial that we understand it. One obvious cause of migration is frontierward expansion – the desire to move into new areas. There's always been a group of new settler who've been willing to do this. Recently 'settled' nations like Australia and the USA are good examples. Secondly there is the rural decline/urban expansion. This refers to the changes brought about by the development of a nation's internal economy: if the rural area suffers recession but the town thrives, people will move there. This can be linked to economic benefits. For example, before the political changes in South Africa many workers came from neighbouring nations to work. Some sent money home, others worked a few months and then returned before making the journey again. Such migration would be seen as economic circulation of migrant labour. Whereas economic migration is voluntary much migration is not. Colonial expansion was a common way of finding settlement for rapidly rising populations (e.g. from the UK and Ireland to the USA and Australia). Migration does not need to be centred on money or force. There are many cases (the USA is a good example) where migration has started because of intolerance at home. Today, there are still cases where people will move to find better social or religious conditions. Finally, there is politics and internal/external policies. Today, no nation exists without some form of migration policy whether it is quite liberal (e.g. Austria) or very restrictive (e.g. Australia). The aim of most migration policies is the development of the national economy and the balancing of internal demands and external obligations. In some countries (e.g. Brazil) internal migration is encouraged so that areas such as the Amazon Basin can be 'opened up' for development. Such political pressures might also build up to create refugees. This is one of the greatest problems of the late twentieth century – the forced movement of people through war or famine. Such forced migration may affect millions of people whether it's the Germans and Russians during the 1930s and 1940s or the civil war victims of Rwanda in the 1990s.

2.1.5 What is population composition and how does it change?

Although major changes in numbers and locations of population are important they're not the only things that demographers need to study. Equally important is the nature of the population: its age and gender for example – the composition of the population.

Two of the aspects looked at first are the gender and age composition. The most common way of displaying such data is through the **age-sex pyramid** (Figure 2.1.9). The shape of the graph can tell us much about the nature of the population. For example, a broad base suggests a high birth rate but if this drops quickly (typical of Third World countries) it indicates a high young-age death rate. The opposite end of the spectrum is the narrow-base pyramid typical of Western Europe. Here, the population is actually falling with old-age people being a far higher proportion. Another aspect is the **life cycle**. Everyone goes through a series of stages each one of which is linked to social and economic changes in our lives. For example, we start as children living with adults. Once we've left this situation we become highly mobile, moving to new jobs and locations. This is often followed by a family which places new demands on social/economic facilities. As children grow another phase is entered until old age is reached. Figure 2.1.10 shows the life cycle and how it relates to age composition.

The life-cycle graph shows one way of describing this phenomenon. Another way would be to look at the statistics of the life cycle: one of the key measures in the **dependency ratio**. This is defined as the number of young and old people as a proportion of the working-age population: $DR = (Py + Po)/Pw \times 100\%$ – where DR is the dependency ratio, Py is the number of 0–15 year olds, Po is the number of people over 65 and Pw is

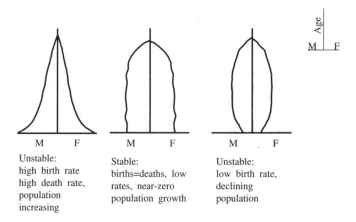

Unstable:
high birth rate
high death rate,
population
increasing

Stable:
births=deaths, low
rates, near-zero
population growth

Unstable:
low birth rate,
declining
population

Figure 2.1.9 Population pyramids

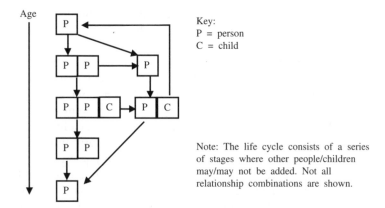

Figure 2.1.10   Life cycles

the number of 16–64 year olds. The figure is of interest to those planning schools, hospitals and accommodation for the elderly.

Another aspect is economic: for example, the census asks questions about jobs and income. What sorts of jobs do people have and how are they changing? What's the unemployment rate? All of these are key questions which it's vital to know the answers to. Often in the UK, census data is divided into 'white' and 'blue' collar jobs (basically, non-manual and manual) and retired.

The last two aspects can be considered together – culture and ethnicity. Any investigation of these areas can be extremely sensitive. It's important to be able to study the cultural and ethic mix in an area because it can be related to a wide range of social issues such as employment, income, social facilities. In many areas of urban UK the cultural and ethnic differences are obvious in the land use and cultural patterns. For example, there are numerous mosques, temples and synagogues in areas where such religions are in the majority. Festivals in schools and elsewhere reflect this diversity. There is also the negative side where social unrest is linked to such areas.

2.1.6   What are the key population issues today?

In this section, a few of the key questions will be outlined and some of the main points they focus on, raised. Given concern about population numbers it is appropriate to start with population growth.

There is much debate about the **population bomb** as one 1970s' book described it. The old idea that the Third World is growing too fast is being replaced with a more considered view of what constitutes a reasonable population growth. If children are the only source of future income, it makes sense to the family to increase the number even if this doesn't work at a national level. What's more interesting is the questioning of the idea of 'zero population growth'. This has been seen as the goal for many nations. Imagine that everybody agreed to have no more than the replacement number of children from today i.e. there would be no increase in population at all. Studies have shown that even if this unlikely event were to occur then we would still be using resources but they would last a little longer. In other words, population growth is not the real culprit but the actual number of people and the amount of resources each person uses. All that zero-growth is doing is giving us a few more years, not a solution. Alternatively, trying to reach low-growth has caused many problems. A common method is birth control. Whilst this is used in many nations in the developed world there are many cases of problems in the developing world. For example, forced sterilisation has led to social unrest in India and China's one-child policy has been blamed for a rise in infant female deaths (where males are more 'acceptable').

What is the ideal population level for a nation? Technically, there are three different levels of population: optimum, overpopulation and under-population. The **optimum** is the greatest number that the area can support at a given standard of living. As the standard increases so the population must decrease which makes this a variable. Alternatively, you could say that a nation can only exist if it can support its own population without outside help i.e. imports. The optimum would be that figure. (this would probably mean a 50% population reduction the UK!). In similar terms, **overpopulation** is having more people than the area can cope with. **Underpopulation** is where the area could support more people than it has. The optimum is fluid because it depends on what standard of living you chose. Overpopulation could suggest a need for aid if local resources fail. Food aid is necessary to help people out but does it make the situation worse because it stops the local market selling food? Underpopulation might be seen as a blessing if we have a crowded world but it too creates problems. For example, in the UK and Australia the working age population is declining. This means that fewer people have to provide more money to help the younger and older members of society. There may well be a limit to this.

Is there a connection between economic development and population? The common argument states that the higher the Gross National Product the lower the growth rate. Thus development is the answer to population problems. The evidence to support this is not clear cut: with poor nations having a low birth rate and the UK's birth rate rising slightly over the last few years! Another approach would be to look not at GNP but at the resources that a nation has at its disposal.

Other population issues deal with the **Ehrlich equation**. Put simply, this tries to relate population to environmental impact: $I = P \times F$ (where I is the environmental impact for a population size of P with a per capita (i.e.

per person) impact of F. In some ways this is similar to the carrying capacity of an area. Thus if either (or both) population or impact increases then the environmental effects will be that much greater. If the world population stayed at the same level (the 'zero option') then the developing nations would still increase their impact because they would presumably raise their standards of living. The alternative would be for developed nations to accept less! This equation also assumes that there's a simple relationship between population and impact; most studies show the effects are far from that simple.

Finally, there is the issue of population policy. Although it differs between nations, this is essentially the government controlling population levels. Most nations have attempted this with varying degrees of success. It's possible to discuss this further by dividing it into two: internal and external policies.

Internal policies relate to the control of the resident population. Birth control is an obvious measure where the majority of nations have some policy. The policy and its effect both socially and on population levels is worth studying. Other internal controls could include voluntary (or forced) settlement. Both Brazil and Russia have used this policy to populate low-population regions. A more negative policy has been the killing of specific groups of the population such as been seen in Russia and Germany in the 1920s/1930s and which is considered to have happened recently in former Yugoslavia and Southeast Asia. External policies usually refer to migration laws. There's often a see-saw relationship as groups are alternatively wanted and excluded. In this respect, Australia provides us with a good example. In the 1950s there was a demand for migrants with the 'Assisted Passages' scheme – a travel subsidy making the journey only £10. By the 1960s this was stopped but migration still allowed. In the 1970s concern was expressed about the numbers who came increasingly from Southeast Asia. Today, there is a very strict quota scheme in operation and, with economic conditions on the decline, migrants have become targeted as 'problems' rather than 'solutions'.

Questions

1. Outline the different ways by which population size can be measured. What are the advantages and limitations of each method?
3. Using newspaper or journal articles, gather a list of migration examples. Note the origin and destination of the people, their numbers and the reasons for their migration. What factors do they share and which are unique to that case? Why should this be so?
3. For each of the life-cycle stages in Figure 2.1.10, what are the implications for migration, housing, and personal income? How would these affect a nation's policies in these areas?
4. Take one nation from the North and one from the South. Using population statistics work on each of their population histories using

the DTT. How closely does the model fit the data? Would it be possible to predict into the future? What limitations are there on this model?

5. Obtain some census data for contrasting regions e.g. NW and SE of the UK. Compare and contrast the economic profiles of the population. Carry out the same analyses for a number of censuses e.g. 60 years. What patterns can you see emerging? How is this related to social and economic policy in the areas?

2.1.7   References

Hornby, W. F. and Jones, M. (1993) *Introduction to Population Geography*, 2nd edition, Cambridge University Press.

Jones, H. R. (1981) *A Population Geography*, Harper and Row.

Ogden, P. E. (1984) *Migration and Geographical Change*, Cambridge University Press.

Sarre, P. (1995) *An Overcrowded World?* Oxford University Press.

Turner, II B. L., Clarke W. C. et al. (1990) *The Earth as Transformed by Human Action*, Cambridge University Press.

Weeks, J. R. (1989) *Population: An Introduction to Concepts and Issues*, Wadsworth.

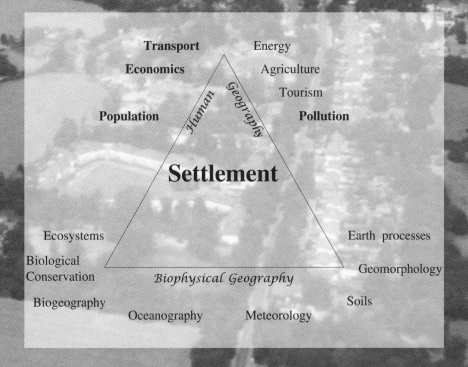

Transport   Energy

Economics   Agriculture

Tourism

Human Geography

Population   **Pollution**

Settlement

Ecosystems   Earth processes

Biological
Conservation   Geomorphology

Biophysical Geography

Biogeography   Soils

Oceanography   Meteorology

Subject overview: settlements

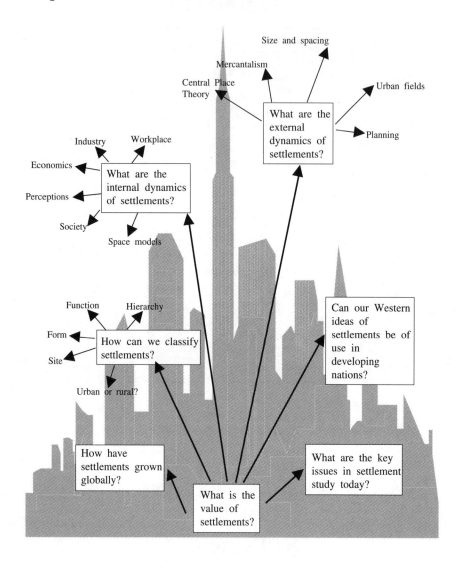

1. The vast majority of people live in some form of settlement.
2. Settlement growth and urbanisation (migration of people to urban areas) has a history stretching back thousands of years.
3. Apart from ancient Rome the world's major cities didn't really take off until the nineteenth century.
4. Settlements can be classified according to their resources and size.
5. Internal dynamics of settlements has led to the creation of numerous models which can be divided into ecological, economic and activity models.
6. Urban image and perception are vital ingredients of how people view their towns and cities. Both can be more influential than any street plan. Neighbourhoods contain areas of like perception.
7. The external dynamics of towns concentrates on size and spacing. Christaller's central place theory is crucial in understanding the relationship although other models are used.
8. Developing-nation settlements have also been central to their nations' growth but theories and ideas from the developed world cannot be transferred as social and cultural differences are too great.
9. Today, problems in settlements such as crime, unemployment and housing have become key political issues.

2.2.1 Introduction

Settlements are one of the key aspects of human existence. Whether this is a small rural village or a mega-city it's still the focus of much of our attention. It's also reasonable to say that because of all the different cultures that use settlements there's grown up a tremendous range of ideas about them. In this chapter we can only hope to look at the most important and point the way forward to others.

2.2.2 What is the value of settlements?

Settlements are not automatic choices – many cultures have nomads and travellers. What are the advantages of living in settlements? One way would be to see settlements as centres of resources (Figure 2.2.1). Here the advantages of settlements can be divided into two: **site** and **situation** (or, **intrinsic** and **extrinsic**). The intrinsic or site features are those things that make the site such a good choice. It might be water supply or a good defensive site or flat land. Around the settlement would be those resources which

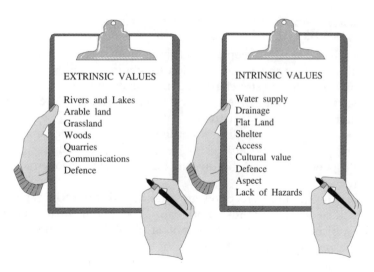

Figure 2.2.1 Site advantages

could be said to be part of the situation of the place (or extrinsic features). These, such as minerals or wood supply would provide the settlement with its basic needs or give it a competitive edge in trade, etc.

2.2.3   How have settlements grown globally?

One of the most outstanding features of human society has been the gathering of people into settlements. When this first happened we can't be sure but archaeologists have evidence going back for thousands of years. Initially, there would be very few people in the settlement but as they succeeded so we assume more people would come from the surrounding areas to the settlement. This idea of settlement growth through migration is called **urbanisation**. We don't know precisely how urbanisation started but several ideas have been considered including agricultural improvements, industrialisation, social/cultural/political sophistication, transportation and natural growth.

> **Tip**
>
> There are three terms which are similar but have completely different meanings. Urbanisation is growth of a settlement through migration; urban growth (sometimes called natural growth) is the expansion of the settlement because of population increase (births); urbanism is the idea of thinking yourself to be urban wherever you live.

Around 1700 there were very few large cities: over 90% of the population lived in rural areas. Some small cities had already started to gain in importance. They would attract people from around the immediate area creating a '**core region**' leaving a **periphery** – a much poorer area with fewer human resources (although their natural resources might be taken to the urban area). Only the Netherlands had enough urban areas to give more than 10% urban dwellers but this seems to have been enough to give them a powerful city-centred economy. By 1800 the picture had changed but only a little. Britain became the next urban-dominant nation. It's been estimated that between about 1750 and 1850 the UK went from being 75% rural to 75% urban dwellers. Moreover, this urban growth was not in one dominant place but in a range of places taking into account the demands of the Industrial Revolution (some geographers have linked this to economic cycles as well, e.g. Kondratiev). With the start of empire, new settlements were founded in colonies which spread urban dominance. Often, these new settlements created 'waves' of human activity moving out from the settlement to surrounding regions involving clearance of forest, drainage of the area, conversion to agriculture and extraction of raw materials. By 1900 the process had spread around the world with remarkably similar results: the dominance of the urban area was virtually complete. However, it wasn't until 1950 that the true nature of urbanisation was obvious with the rapid growth in developing nations.

Such human concentration can't take place without some impact on the area whether it is environmental (e.g. pollution), economic (e.g. concentration of wealth in cities) or social. Here, the desire for more living space is shown by people moving out of towns, **counterurbanisation** or of their possible moving back (**counter-counterurbanisation**). Recently it has also been possible to detect a further trend, when people move back into inner-city areas for the convenience and lack of travelling often restoring older run-down areas in the process (known as **gentrification**).

2.2.4 How can we classify settlements?

Settlement classification is important. It allows geographers to group like settlements and then to compare and contrast different groups. It can also help to explain why certain areas have higher settlement concentrations than others. There's more than one set of criteria for classification and each one has its advantages and disadvantages:

1 **Site and situation**. The site refers to those qualities which make the actual place worth building on. For example, it might have a good water supply such as you find in the spring-line villages along the chalk/clay boundary in Southern England. Other examples would be relief, defence, bridging point across a river and hazard avoidance. The situa-

tion refers to the resources/facilities around the site and so good farm-land, fuel supply and mineral resources are all good examples.

2 **Form**. The shape of the settlement can give clues to its origins. So, for example, rural settlements can be classified according to whether they are isolated (far from other settlements), nucleated (centred around a point), dispersed (spread out), ribbon (forming a long, thin line) or planned. Although a useful criterion it's worth noting that it's not used too much today apart from looking at historical geography and rural settlements.

3 **Function**. Just as the form is less used, so function is an increasingly important classification. As the name suggests it is a criterion based on the need for the settlement – what is it doing there? Functions can vary from nation to nation and between those settlements in mainly rural or urban settings. Examples of function would include market, transport point, defence, dormitory/satellite, resources, services, trade, religious/cultural, recreation, political and administration.

4 **Hierarchy**. A hierarchy is a grouping based on size or importance. Starting with hamlet it moves through village, small town, large town and city to metropolis (or world-city – a place of international importance).

5 **Urban** or **rural**. This final classification might seem very simple but it is actually very powerful; numerous geographers have spent time examining this question because it's linked to so many other aspects of human life. For example, a study carried out in North Hampshire in 1966 wanted to see how village inhabitants viewed their places. Despite the 'rural' nature of much of the county at that time (the major planned settlements had yet to be built) the survey showed that most people were connected to the town not just because of job but by perception and attitude. The survey called this '**mental urbanisation**' and it does have important consequences. For example, how far should villages go in trying to compete with town facilities? Or, imagine trying to provide a rural doctor service. If most people were focused on the nearest town how would one view the staffing of a village surgery?

2.2.5 What are the internal dynamics of settlements?

Towns and cities are dynamic, complex places: models aid our understanding. Although the range of topics can be very broad, in practice, five aspects are seen as most important: space models, urban social geography, perceptions, economics, industry. The focus here is on one type of settlement: the developed world urban area. Rural areas lack the dynamics and, often, developing world urban areas are less well researched.

Space models are amongst the most famous geographical ideas. They have been devised to help explain how a settlement grew and why it contained the divisions (residential, retail, industry) that it did. Most models fall into one of three approaches: ecological, economic and activity system. **Ecological models** are based on the study of human ecology. A group of 1920s' University of Chicago sociologists produced the first models. The key idea was one of quality of land and competition to pay for it. They could see from their own study of Chicago that there were definite areas but it was Burgess and Park who put forward the first model – **the concentric zone model** (Figure 2.2.2A). The most expensive land would be the most accessible and therefore this would be the central business district (because these businesses could afford it). Outside that, poor housing/slums would be occupied by the poorest residents. Successive

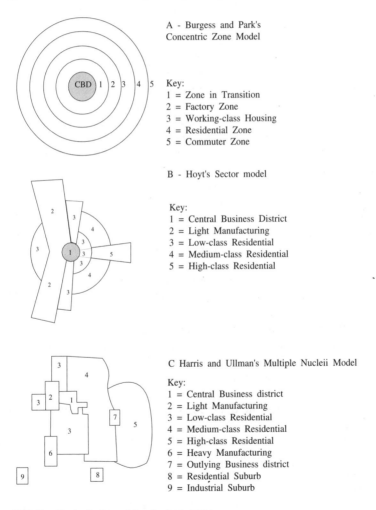

A - Burgess and Park's Concentric Zone Model

Key:
1 = Zone in Transition
2 = Factory Zone
3 = Working-class Housing
4 = Residential Zone
5 = Commuter Zone

B - Hoyt's Sector model

Key:
1 = Central Business District
2 = Light Manufacturing
3 = Low-class Residential
4 = Medium-class Residential
5 = High-class Residential

C Harris and Ullman's Multiple Nucleii Model

Key:
1 = Central Business district
2 = Light Manufacturing
3 = Low-class Residential
4 = Medium-class Residential
5 = High-class Residential
6 = Heavy Manufacturing
7 = Outlying Business district
8 = Residential Suburb
9 = Industrial Suburb

Figure 2.2.2 Ecological models of urban areas

zones would have better quality housing. The city would grow by an influx of residents (who, being poorest, would live in the centre) and the displacement of existing residents to better quality housing (much like invasion and succession in ecology). Despite numerous criticisms (some, e.g. limited land use studies, were valid; others, e.g. it was based on only one place, less so) this model has been seen as the basis for other work either by refining original ideas or creating new models. This latter option was tried first by Homer Hoyt in his **sector model** (Figure 2.2.2B) and then by Harris and Ullman in their **multiple nucleii model** (Figure 2.2.2C). Hoyt accepted that the zones would rarely be found in real cities and that a linear approach (to take account of transport lines for example) would make a better model. Harris and Ullman added the extra idea of having more than one growth point. Other researchers tried different sets of factors in different areas. In a study of Sheffield, Mann used a combination of zone and sector models in an attempt to explain land use variations. Robson's work on Sunderland used three factors – socio-economic status, age structure and housing environment trying to add more variables for a more realistic picture, whilst Rees put in ethnicity, racial origins, as a variable (see Figure 2.2.3). Most of these ideas used Burgess and Parks' original concepts. There have been other ecologically oriented models which have used different approaches. Three US sociologists – Shevky, Williams and Bell – used census data to provide detailed maps of a city. They mapped the distribution of social characteristics – socio-economic status, family structure and race – to produce a **social area analysis model** (later refined with statistical techniques to become **factorial analysis** and used by Murdie in his study of Toronto).

In 1903 an economist, Richard Hurd, published the '*Principles of City Land Use*' which said that land use patterns are based on land values and started the first of the **economic models**. The nearer to the centre of an urban area, the greater the land value. Take a specific land use like retail shopping. Stores will pay high rents because they need the accessibility. This will produce a graph (the so-called bid-rent curve). Housing will pay less to be in the centre (it doesn't really need the access) but its curve will drop off far more slowly. By adding all these different bid-rents together the bid-rent model of urban land use can be devised (an example of which is Figure 2.2.4). The central idea is that the land use producing the greatest profit will dominate any given area.

Activity models assume that people do not follow blindly some ecological or economic force but have their own ideas. The origins of this perspective are not well recorded but we do know that as long ago as 1936 planners looking at London's growth considered it to be of mental rather than economic origin. In other words London was growing because people thought it was a good place to be rather than for any actual evidence.

These space models have something in common. They consider the 'space' part to mean horizontal space (much like a flat map). There are two

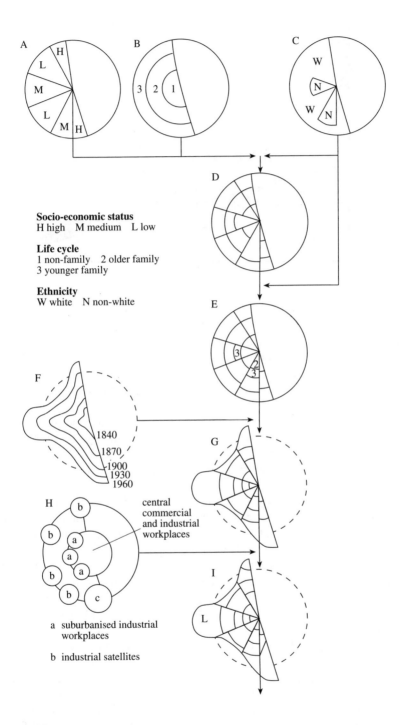

Socio-economic status
H high   M medium   L low

Life cycle
1 non-family   2 older family
3 younger family

Ethnicity
W white   N non-white

central
commercial
and industrial
workplaces

a suburbanised industrial
   workplaces

b industrial satellites

Figure 2.2.3   Rees' model

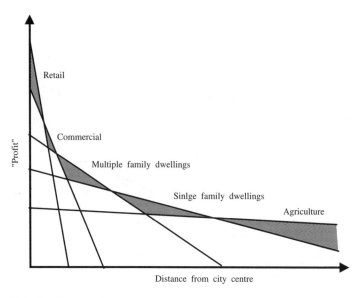

Figure 2.2.4 Bid-rent theory

other approaches which are worth looking at. The Australian economist, Colin Clark, examined a range of cities and noticed that whatever else the differences there was one point of agreement – the population density declined away from the centre. Also, it would appear to alter through time as the city grows. This work was extended by the American Brian Berry (Figure 2.2.5). The other approach used the change in building height. In the centre you would find the tallest buildings because ground rent would be so high. As you went from the centre so the building height would decline – it would be cheaper to build outwards rather than upwards!

Urban social geography is an important area of study because it covers some of the most sensitive relations between people (for example, riots and crime as well as community). The so-called inner city problems of London, Liverpool and Bristol which have attracted so much media and political attention over the last twenty years are examples of this. Much of the work in this area tries to explain what people can be found where and how they view their surroundings. For example, what causes people to locate in similar groups, i.e. **segregation**. Is it defence, avoidance (by recent migrants), preservation (of cultural identity) or attack (strength in numbers)? William Firey argued that the '**urban image**' was important. Everybody had their own idea of what constituted their place. He called this area social space which would be made up of activity space (your job and its surroundings) and awareness space (home territory). In this sense, an urban area would be a collection of thousands of individual social spaces! This individual nature would come about because the factors

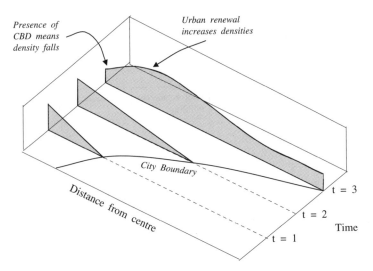

Presence of CBD means density falls

Urban renewal increases densities

City Boundary

Distance from centre

t = 3

t = 2

t = 1

Time

Figure 2.2.5   Density gradients in Western cities

affecting social space: culture, prejudice, motivation and psychology were unique in every person. From these ideas people have made the concept of **neighbourhood**. Usually, they form through social interactions over many years but recently, planned neighbourhoods in New Towns have been attempted (⇒ planning) but rarely with any success.

The third aspect is **perception**. Much of what Firey was writing about had to do with people's perceptions of a place rather than the reality of the land use. The best way to demonstrate this is with a simple experiment (a mental map).

The point is that people have a different perception of their area. What might matter more than the actual land use is what people *think* is the land use. If they define one area as their community and yet another area is actually mapped out then there's little chance that the mapped community will be successful. In one study of Los Angeles' children the mental maps of poor residents were just a few streets whereas the rich children saw the whole city (and especially the freeways) as part of their community.

The fourth aspect is **economics**. With the concentration of population and therefore employment, the economics of settlements is going to be an important factor. One of the earliest economic uses of the settlement would be as a market. Here, trade would bring money into the settlement which, according to Allan Pred, would lead to growth in both the settlement and its urban income, a cumulative process (Figure 2.2.6) – success breeds success! Other writers have argued that while this does happen, the effects are not equal. Growth would tend to concentrate in large towns because they have the size for a mixed economic base, a built-in market for goods, political muscle and the opportunity for innovation diffusion (e.g. M4 Corridor from London to Newbury as a 'high tech' centre).

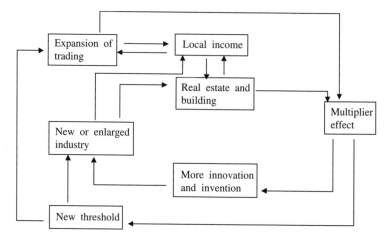

Figure 2.2.6   Circular and cumulative causation

The fifth aspect is **industry**. Considering its importance there's been less written than you might expect about its position in settlement land use. Burgess and Park tended to consider residential land use as important although they did describe both the central business district (CBD) and an industrial zone. Harris and Ullman's multiple nucleii could well have one of the nucleii as a mine or industrial centre drawing in a localised population.

2.2.6   What are the external dynamics of settlements?

Although the internal dynamics of a settlement are crucial they cannot be considered in isolation. One of the key aspects of urbanisation is the relationships between settlements. Walter Christaller published, in 1933, his study of the distribution of settlements in Southern Germany. From this has grown the idea of **central place theory**. Imagine a flat, uniform landscape where all resources are evenly distributed (actually called an isotropic surface). People on that landscape will use its resources and live in settlements. The spacing of these settlements will be even (because we assume all settlements have identical numbers of people). It follows that the land that they use will also be equal in area. If you try to put this on a map, the only pattern that fits with no space left over is the hexagon. At this stage we have a number of villages in the centre of their hexagonal fields (why central? – because of ease of access; why hexagonal? – because Christaller was working on a theoretical model and it's the only shape that uses up all the space!). Assume that this area is going to develop. Which settlements would be most likely to have the advantage? According to Christaller, the one in the centre (see Figure 2.2.7). This leads to some key concepts:

1 **Centrality of a place**. As an area develops the most important settlement is the one in the middle because it has the best ease of access (and therefore the biggest market share) compared with other settlements.
2 **Sphere of influence**. The area from which people come to go to a particular settlement. This can also be called the urban field (see below). This will vary with the size (and therefore importance) of the settlement.

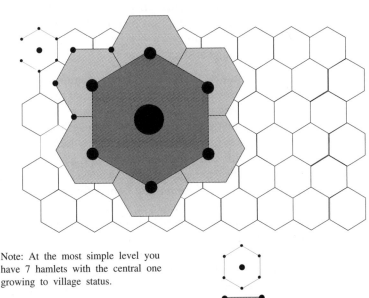

Note: At the most simple level you have 7 hamlets with the central one growing to village status.

In the same way, the central village can grow to be a small town.

Again, the central small town will have a sufficiently large threshold population to grow to be a large town.

In this diagram, only sample settlements and areas have been marked. Remember, as you increase the hierarchy so the threshold area (or sphere of influence, gets larger.

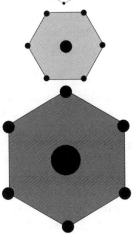

Figure 2.2.7 *Christaller's central place theory*

3 **Threshold population**. The number of people needed to make the provision of a service economically viable. Thus for everyday things like food, the threshold population is small (everyone needs food). However, for a jewellers, the threshold population is larger because you don't need jewellery every day.

4 **Hierarchies**. The whole idea of Christaller's model is that it suggests a series of urban hierarchies: village (at the bottom), small town, large town, city and international centre (at the top). There's also the idea that functions (something that we need/use such as food, doctors etc.) have their own hierarchy – low order functions such as food shopping to high order functions such as international banking. Higher order functions need a higher order settlement because they need a larger sphere of influence. This idea of function hierarchy has even been used to distinguish between settlement types. For example, jewellers seem to mark the line between village and small town.

In Christaller's area of Rheinland Pfalz, the planners were using his model to plan the distribution of growth in their area (only using time rather than distance as the factor). For example, sports facilities were given a particular threshold population and time (30 minutes). Thus all sports centres were designed to be no more than 30 minutes from any home.

One of the problems with central place theory is that it assumes that growth is generated internally. Another is that it takes no account of historical development. What happens if something comes form another region or nation over a period of years? One way you could get an external influence is from (merchants) trading. Richard Vance Jr. used this idea when he developed his **mercantile model** (Figure 2.2.8). You start with the initial search for trade. If sufficient produce is found then the trade, based on primary products, food and timber etc. can start using the port (or gateway city as it's called). Soon settlers will arrive and gradually colonise the interior (or hinterland around the port). As this develops so internal trade will be started. The final stage is when you get hierarchies developing in the nation as a response to things like population growth and economic development.

One feature common to both models was the idea of size. G. Zipf proposed the **rank–size rule** in the late 1940s. The largest city in the nation has a size S and a rank of 1. According to Zipf the second largest city would have a size of S/2. This can continue so that the nth ranked settlement has a size of S/n. Trying to fit actual nations' figures gives interesting results. There appears to be a difference between developed and developing nations. In some countries, the biggest city is so large it dominates and the next biggest is not S/2 but maybe S/6. We call this extra-large city the **primate city** (or **binary cities** if there are two close together like Sao Paulo and Rio de Janiero). Although the primary city might seem a good idea it can be too large and can suck in resources leaving surrounding areas under resourced (Paris is a good case here).

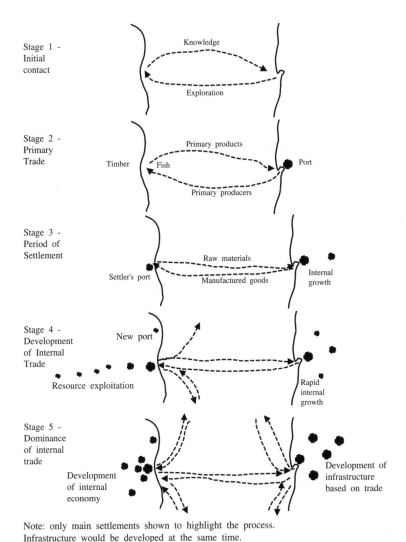

Stage 1 -
Initial
contact

Knowledge

Exploration

Stage 2 -
Primary
Trade

Timber

Primary products

Fish

Port

Primary producers

Stage 3 -
Period of
Settlement

Raw materials

Settler's port

Manufactured goods

Internal
growth

Stage 4 -
Development
of Internal
Trade

New port

Resource exploitation

Rapid
internal
growth

Stage 5 -
Dominance
of internal
trade

Development
of internal
economy

Development of
infrastructure
based on trade

Note: only main settlements shown to highlight the process.
Infrastructure would be developed at the same time.

Figure 2.2.8 The mercantile model

Central place and rank size are useful on a national scale but in a local context two other ideas are useful. The **centrality index** is a ratio of the number of a functions the town has compared to the number found in the region. Thus:

$$C = (t/T) \times 100$$

where C is the centrality index, t is the share of a function the settlement has and T is the total amount of that function in the area.

The break-point theory (or Reilly's model) is a way of working out where the boundary between two settlements could go. The line separating those who go to town A as against town B can be calculated:

Break-point from A = distance of A from B/1 + $\sqrt{pA/pB}$

2.2.7 Can our ideas of settlements be used in developing countries?

Considering the size of the developing world and its importance the amount of material produced on settlements is minute in comparison to that found for the developed world. Models which explain the dynamics of cities in the North do not cover the range of factors experienced in the South e.g. **colonialism** and trade.

Examples of these differences can be seen when you try to compare space models like Burgess' with those made for developing world cities e.g. compare Figure 2.2.2 and Figure 2.2.9. Turning to the social geography of developing-world settlements one major difference can be found which is commented upon by most writers: the existence of the 'informal' economy. This is usually taken to mean that part of the urban income which does not go through official/taxation channels. It's seen in every country but there does appear to be some evidence that it's a greater part of the developing nation urban income than usually found in developed nations. In the past such action was disapproved of and there were schemes to try to remove it. Of course, these failed and today there is growing acceptance to see it as a legitimate part of the growth of settlements. It's seen as one of the number of ways by which people can express themselves in terms of economic/social/political options.

2.2.8 What are the key issues in settlement study?

Much of the concern focuses not so much on the theory of settlements but on the practical issues of community and social relations. The idea that many problems in settlements are, if not caused by, at least made worse by, differences in equality is central to much study. The aim here is to highlight some of the more important issues:

1 **Inner cities**. The so-called 'inner-city problem' has been with us for decades but recently it's been given more attention through a series of riots (which in the UK were in Toxteth, Liverpool and Brixton, London in the early 1980s). The reaction to these events was, in part, a geographical one. There was a link made between the distribution of

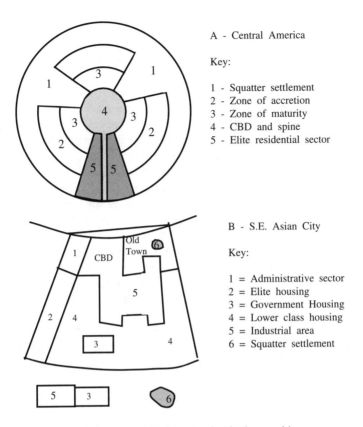

A - Central America

Key:

1 - Squatter settlement
2 - Zone of accretion
3 - Zone of maturity
4 - CBD and spine
5 - Elite residential sector

B - S.E. Asian City

Key:

1 = Administrative sector
2 = Elite housing
3 = Government Housing
4 = Lower class housing
5 = Industrial area
6 = Squatter settlement

Figure 2.2.9   Burgess' model applied to the developing world

social disadvantage and crime. Thus planners were called upon to 'design out' crime from some of the more decaying areas. Just as with the problems of towns in the late nineteenth century. ($\Rightarrow$ planning) it was considered better to plan than to tackle the injustices felt by the community. Inner-city problems are not just about crime. The changing nature of industry ($\Rightarrow$ economics) has forced many businesses out of the inner-city areas leaving high unemployment behind.

2 **Out-of-town centres.** The economics that led to the development of the high street also led to its failure. High land prices and planning problems for expansion meant that it was often better to relocate on the outskirts of an urban area. The result would be a decline in the high street shops (which could spiral downwards as more shops closed) while the new centres were booming. Some places, such as the Gateshead Metro centre have generated considerable income but at the expense of traditional areas. Today, there are moves to reduce this trend as the out-of-town centres are seen as being wasteful of space.

3 **Rurban fringe.** This area, halfway between rural and urban areas is a

source of considerable debate especially in areas where there is pressure to develop. In some ways this goes back to the old ideas of urban sprawl in the 1930s. This took up farmland (often of higher quality than elsewhere because the early settlers would locate near the best fields). Planners are concerned to see such land be used to best advantage whereas developers and landowners want to see the best return on their money. Much of the debate is quite strong especially in the Southeast of England where the pressure to develop is greatest. Ironically, what was once seen as a welcome addition to urban income is now seen as an unwanted population increase where the ideas of NIMBY (not in my back yard) often sum up residents' concerns.

4 **Central business district (CBD)**. This is an area of office and commercial development which can be seen as vital to the life of the area. Not only is it the most accessible area in the settlement (which means it commands the highest rents) it can also provide a number of vital services including retail outlets, offices, specialist areas (such as Hatton Garden in London – a jewellery centre where aggregation of shops leads to the benefits such as cumulative attraction) and vertical zoning (distribution of functions by storey). Because of its importance many people have studied the area trying to analyse its activities. We can see the CBD as a key issue in the continued development of the town.

5 **Rural settlements**. At the other end of the settlement scale we have the small villages and hamlets of our rural areas. Given the movement of population to the towns there are often too few people left to justify the provision of major social and community services such as doctors and buses. The difficulty is not just in these 'remote' areas: there's also a difficulty for rural areas on the edge of towns. Here, rather than have too few services they can have too few rural workers as they can't afford the house prices in the area. The issues of the more remote rural areas in the UK are, in many ways, similar to those of the South where excess rural depopulation has led to serious economic problems for their regions.

There are two points to note with these brief examples. Firstly, the problems outlined are, in reality, highly complex involving a very wide range of people and issues. There's no such thing as a purely settlement problem. Secondly, geography can pay a large part in both understanding and resolving the problems. Thus our knowledge of the theory of settlements seen in previous sections can now be put to good use in discussing and tackling some of the most critical social issues today.

Questions

1. Using a map of your local area chose 12 settlements at random. For each one, note the intrinsic and extrinsic features of the site. Are there are aspects in common? Do some sites have particularly good features? How can you explain this?

2. Take a large piece of paper (A3) and a pencil. Allowing yourself just two minutes, draw a map of your town (or rural area). Get some friends/colleagues/adults to do the same thing. Compare results. What do you see?
3. How do you think that these different outlooks on community (as shown by 'mental mapping' would affect the social cohesion of the area?
4. Try to construct rank-size graphs using a variety of nations. What patterns can you find? How could you explain them?

2.2.9 References

Carr, M. (1987) *Patterns: Process and Change in Human Geography.* Macmillan.

Carter, H. (1995) *The Study of Urban Geography,* 4th edition. Arnold.

Daniel, P. and Hopkinson, M. (1989) *The Geography of Settlement,* 2nd edition. Oliver and Boyd.

Dickenson, J. et al. (1996) *A Geography of the Third World,* 2nd edition. Routledge.

Hornby, W.F. and Jones, M. (1991) *An Introduction to Settlement Geography.* Cambridge University Press.

Knox, P. (1994) *Urban Social Geography: An Introduction.* Longman.

Turner, II B.L. et al. (1990) *The Earth as Transformed by Human Action.* Cambridge University Press.

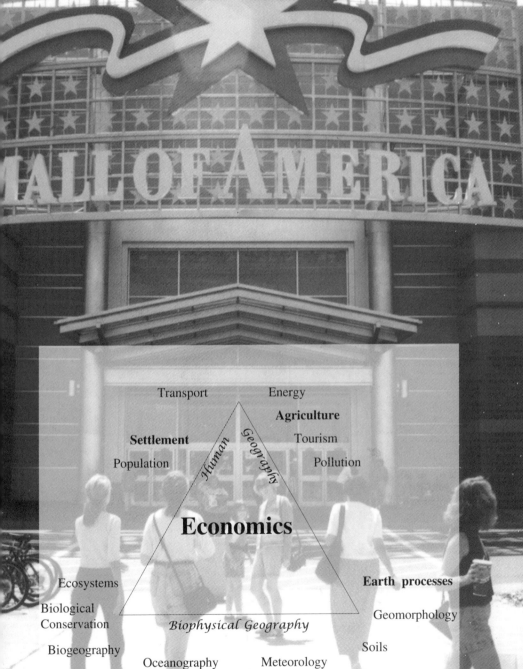

Transport   Energy

Agriculture

Settlement   Tourism

Population   Pollution

Human Geography

Economics

Ecosystems   **Earth processes**

Biological
Conservation   Geomorphology

Biophysical Geography

Biogeography   Soils

Oceanography   Meteorology

Subject overview: economic geography

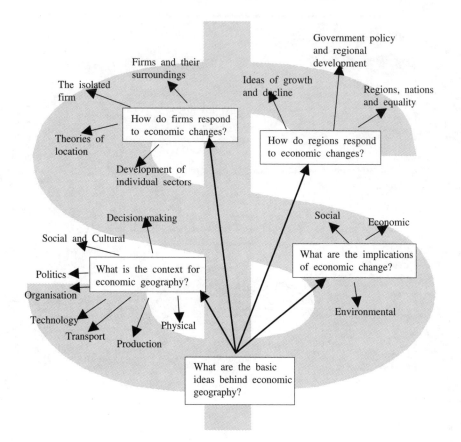

1. With the changing structure of industries and national economies it's important to see the geographical implications of this.
2. Economic geography has been divided into studies of the individual firm and the region.
3. Context is important. Numerous factors shape individual decisions.
4. Economies seem to go through a series of wealth-producing stages as sectors from primary to quaternary take over as key sectors.
5. Economic models such as Weber and Losch have attempted to explain location in terms of supply and demand mathematics.
6. Regions change through the sum total of their industries. Common models assume a cumulative factor.
7. Most governments have tried to affect regional growth through positive (growth) and negative (restriction) policies.
8. Changes in economic structure have social, economic and environmental significance.

2.3.1 Introduction

The past decade has seen some of the most fundamental changes in the economies of people and nations. The whole structure of the national economy has been altered. No longer is manufacturing the key sector but services and information technology lead the way. The whole idea of the national economy is under question as multinational corporations work outside traditional boundaries. At the personal level, work is changing its structure with increasing participation by women and a large increase in part-time and contract work. Even the 'certainties' of the mid-1980s, e.g. the rise of Japan and Asia have been confounded by the financial problems of the mid-1990s. To make sense of this one needs to look at the realm of economic geography.

2.3.2 What are the basic ideas behind economic geography?

The answer to this in one sentence: the study of the generation of wealth by economic sectors occupying various locations all of which change through time. Most economic activity involves the exchange of some item (whether it's food, machinery or even an idea) for money. The aim is to make more money than you had to pay in the first place. This implies that we are dealing with a capitalistic society where nothing is provided unless a profit can be made. These two ideas are crucial and underpin most of the work described in this chapter.

Commonly, economic activity is divided into four broad sectors: **primary** industry (farming, fishing, forestry and mining – the extractive industries), **secondary** industry (manufacturing), **tertiary** (services) and **quaternary** (banking and information services). Some writers also mention a **quinary** sector which would be information services but this is not universal. The relative importance of these sectors changes through time. The Clark–Fisher model (Figure 2.3.1) argues that as a nation develops so the proportion of its wealth produced from each sector changes from primary to secondary etc. This also implies that the employment structure in each sector will change as well, e.g. the UK farming population in 1921 was 996,000 and only 488,000 in 1966.

2.3.3 What factors affect location?

Firms are there for a reason. This might change between different companies in the same sector and will certainly change between sectors. The aim of this section is to outline the range of potential factors that could be involved in locational decision making. They can be divided into three main groups: physical, production and organisational. Although the traditional focus has been on the manufacturing sector it is just as possible to focus on the service sector.

Physical factors are of fundamental importance to both primary and secondary industries e.g. coal mining, There's an optimum location i.e. where the seams are largest or the conditions most favourable. If that were the only place then there would be little activity. Around the optimum is a sub-optimum i.e. good but less than perfect. Whether this is used depends on a series of factors such as cost of production, demand, supply from elsewhere and the technology available. It follows that this sub-optimal zone is not fixed but depends on how it's perceived (i.e. it's a human rather than physical limit). Allied to these factors are **transport factors**, traditionally a key factor in location. For example, the distribution of market gardens around London was dependent on the transport getting the produce there still fresh. Today, this need has been removed – fresh food can come into the UK from all over the world. However, transport *costs* are still an important consideration ($\Rightarrow$ transport). Finally, there are **technological factors** – probably the most important single item today. The development of new high technology industries has opened up considerable changes in industrial location. Two concepts are worth exploring: Kondratieff (see below) and product cycles. The product cycle is a series of stages through which a product is said to go from invention to decline. Although this can refer to

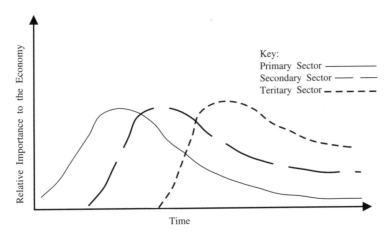

Figure 2.3.1 The Clark-Fisher model

a single product such as a specific model of washing machine, it can also refer to a firm or even an industrial sector. At this level, a study of the product cycle becomes increasingly important in understanding locational changes. Figure 2.3.2 shows that the product life cycle has been linked to both employment and the location of manufacturing.

Secondly, **production factors** cover those aspects of direct concern with the business: land, labour, capital and demand (the market). **Land** can be a complex aspect of economic location. It depends on how one views the idea of land: a resource, a place or an example of power all influence decisions about it. Land can also be seen as an investment. Much farming land was subject to purchase by pension fund investors as a way of making money in the UK in the 1980s. The collapse of the land market in the late 1980s had (and continues to have) serious implications for businesses. Whatever the employment characteristics of each sector, **labour** is still a crucial factor. Every firm needs to consider the type (gender, age, skills), availability (sufficient labour for investment to be worthwhile), mobility (can workers afford to move to jobs) and cost of labour. As technology changes so do requirements for labour leading some writers to propose a four-fold classification of businesses and their labour supply: manufacture (where you bring workers to a specific place); mechanisation (some labour replaced by machinery); Fordism (the rise of specialised assembly-line techniques) and neo-Fordism (the replacement of some labour with automated devices). Without **capital** (i.e. money and assets) firms cannot really compete. Although traditionally the flow of money has been seen as fluid i.e. with few restrictions, in reality capital tends to be subject to numerous constraints including access to finance, inertia (fixed assets are not easily transported), government policy (especially regional policy where grants/subsidies may be available), existing versus new and lending

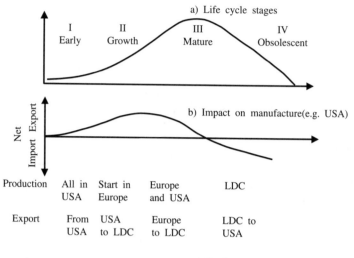

a) Life cycle stages

	I	II	III	IV
	Early	Growth	Mature	Obsolescent

b) Impact on manufacture(e.g. USA)

Production	All in USA	Start in Europe	Europe and USA	LDC
Export	From USA	USA to LDC	Europe to LDC	LDC to USA

c) Production Characteristics

Stage	I	(+)	II	III	IV
Production	Short Run			Mass Production	Long Run
Capital	Low			High	High
Structure	Many firms			Declining firms	Few firms
Labour	R&D			Management	Unskilled

Figure 2.3.2 The product lif cycle

policies of banks. The level of **demand** is dependent upon the level of economic activity in the area. The greater the activity, the greater the wealth and so the higher the level of expenditure (i.e. demand). If demand for a particular product is high enough then an area might have its own firm dealing with it. The idea that a certain level of demand is required before a service is provided is called the **threshold** for that service which might vary from a few hundred for a corner store to many tens of thousands for a large supermarket. Of course, firms don't act in isolation. There are two instances where combinations of firms can be seen to be to the benefit of all: **linkages** where firms are joined in terms of providing goods and services to each other (e.g. a car plant getting supplies) and **agglomerations** where a group of similar firms locate near each other to generate more business (e.g. jewellery firms in Hatton Garden, London).

Thirdly there are those aspects which can be grouped under the heading of **organisational factors**. Within this heading there is the **structure** of the company. There are basically four ways for larger companies to organise: by function (e.g. finance, production etc.); product or service (e.g. product a, product b); territory (e.g. place a, place b) or by holding company (e.g. company a, company b) where in each case the unit in the brackets represents a different location. Secondly, there is the degree of **integration**. This refers to the way in which the larger company relates to its units. For example, many agricultural operations buy firms dealing with all aspects of food production from farm to table (called **vertical integration**) while others try to dominate a particular sector (**horizontal integration**). Hymer has argued that larger companies are formed into **hierarchies** and that the location of a unit depends on the position in the hierarchy. Thus head offices are in major cities, general administration is found in regional centres leaving production to peripheral areas. Lastly there is **rationalisation**. One of the advantages of integration is that it can save money i.e. costs are lowered because there is the possibility of some economies of scale. Thinking along these lines means that five separate companies have higher costs (in administration etc.) than a five-unit company. On similar lines it makes sense to concentrate economic activity at special units rather than have duplication.

Although these factors are internal to the firm there are other, external ones. For example **political factors** refer to the national and international institutions that govern the way in which business etc. is carried out. This can be seen in two ways: legislation and government policy. **Social and cultural factors** are noted because economic activity is not just about firms and their locations. It's also about the type of people that work and live in that area. Because the social and cultural values in an area change over time it's impossible to truly understand an area unless you know about its history i.e. location finding is not a neutral exercise. If economic and social systems are so integrated then it follows that any change in one will be seen as a change in the other. Lastly there is **decision making**. In all of this we have often assumed that the decision-maker has the abilities of a 'rational economic person'. What this means is that the person taking the decisions has all the information possible to decide about a location (called perfect information) and that the person only acts in accordance with strict economic principles. Of course such situations are only found in models. Normally, other aspects come into play. For example, capital is invested in a firm. This makes goods which are sold at a profit. The profit goes to buy materials to make even more goods (called **circuits of capital**). Thus a location which allows this to occur most profitably would be the one chosen. Alternatively there is the **satisficer** concept. This recognises that all people have imperfect knowledge – there's always going to be some detail they don't know. However, they make a decision based on the best available knowledge (which is, of course, dependent upon their age, gender, experience etc.) which will give a satisfactory result. There's no reason why the only decisions are related to the firm or economic decisions. Personal preference may well play a part (a desire to be near a specific place or a golf course or shopping area etc.).

2.3.4 How do firms find their sites?

The Clark–Fisher model demonstrated the changes that take place through time within sectors. As sectors change so will the locational forces upon them. If we accept that these changes will occur then the next question is how. Both Taylor and Watts have described two ways: **merger/rationalisation** and **market area extension**. In merger and rationalisation, one company takes over another one (merger) and uses the benefits (economies) of scale to close some sites (rationalisation). This results in fewer, bigger sites. The other possibility is market area extension. Imagine a situation where there are several sites of different sizes each producing the same product for different sized markets. The larger sites have the advantage of economies of scale. If their costs are lower they can afford to pay more for transport and still undercut the smaller sites nearby. These go out of business and the larger site grows even bigger. Repeat this often enough and you have a handful of sites producing for a very large area. These ideas relate to changes through time. Others relate to changes due to location. Many economic geographers start out by considering the initial location of the firm.

Much of the work on location has come to us in the form of model building. The factors used in these models can help highlight those seen as most important. One of the first, and still one of the most widely cited is that devised by Alfred Weber. In his 1909 text '*Theory of the Location of Industries*' he devised his classic location model (see Figure 2.3.3). Assume that there is one source of raw materials and one market (and all other factors are evenly distributed). Weber argued that the firm would locate at the place of least cost. To find this he drew a series of circles from the source and the market. These isotims would be lines of equal transport cost. At each point where the circles from one cross the other it would be possible to work out the total transport cost. By using these points to draw another set of lines (isodopanes) you can find the least cost.

For Weber, location was a question of mapping the differences of the cost of producing and delivering a product. Losch argued that what was important was to maximise the profit. The best location would be where the sum 'sales – costs = profit' was at its greatest. Plotting this would produce a 'demand cone'. For Losch, location was market-oriented rather than production-oriented. Other researchers have taken a slightly different approach to this topic by using a range of costs, substitution of one good for another etc., but the basic idea remains the same.

These models have all used the same basic concepts: the reduction of costs and the maximisation of profit (which is much the same as most firms try to achieve). Not all the models take such a strictly economic view. These ideas, not always obvious, extend our appreciation of how location might work in real-life situations. For example, Weber's 'uncertainty model' bases its ideas on the fact that no one has perfect knowledge. The

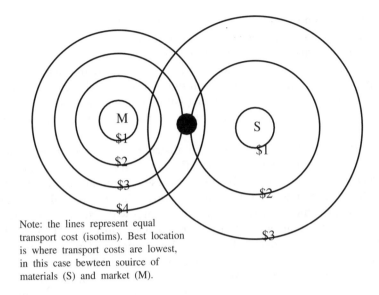

Note: the lines represent equal transport cost (isotims). Best location is where transport costs are lowest, in this case bewteen souirce of materials (S) and market (M).

Figure 2.3.3 Weber's location model

greater the uncertainty, the smaller the site. Risky ventures, distance from the market, price variability, distance from supplier etc. all contribute to this effect. The result is that newer firms with poorer knowledge would locate near the urban areas to obtain a market and any services they lacked. Other firms might locate nearby for the same reasons and so you'd end up with an agglomeration of businesses near urban areas. Alternatively, Hotelling used competition as an explanation for location. His classic example used ice-cream sellers on a beach (Figure 2.3.4). They locate at an equal distance for their half of the beach (i.e. at the quarter and three-quarter points). If one wants more territory then it makes sense to moves close to the other's place. If this is done, then the other seller leapfrogs over

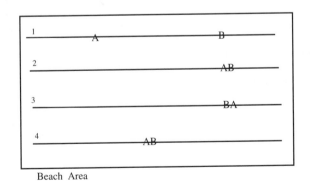

Beach Area

Figure 2.3.4 Hoteling's ice-cream seller model

the first seller and so on. Eventually, each will end up, back-to-back at the half-way line. Working this out in Weberian terms gives some interesting results. In the first place (the 1/4–3/4 siting, the average costs for each customer are at a minimum (no one needs to travel further than 1/4 of the beach). This would be the best solution for society. By locating in the middle, the average costs are higher (with almost half a beach to walk for some people) but this is the best place for the sellers in terms of market forces.

Models based on ideas of even distribution of resources, rational economic people etc. have their benefits but with today's complexity they are seen to have limitations. To counter these mathematical models, other ideas have been suggested. Greenhut distinguishes between those factors which are the same everywhere (e.g. labour rates with a trades union agreement) and those that vary with location (such as transport). His three-fold locational divisions (demand, cost and personal preference) provide a detailed alternative. In a similar fashion, Townroe has a set of location factors to be taken into account by people setting up firms (Table 2.3.1 compares the two ideas). Pred takes another perspective in his behavioural model. Here, the variables are information and the ability to use it. Put this as a matrix and you'll find everybody will go somewhere on it depending on their knowledge and ability. Those with better knowledge and ability will locate closer to the optimum location.

Whichever model is chosen there is still the same fundamental criticism to be made: no firm is isolated because every firm needs someone else to provide at least some of the items it uses or to give it some advantage. **Agglomeration economics** describes the benefits to be gained by prox-

Table 2.3.1 Location factors

Factor	Greenhut	Townroe
Government policy		✓
Transport and communications	✓	✓
Labour	✓	✓
Markets		✓
Raw materials	✓	✓
Services		✓
Local government services		✓
Amenity	✓	✓
Inter-urban locations		✓
Physical factors	✓	✓
Tenure		✓
Buildings		✓
Infrastructure		✓
Price	✓	✓
Competitors	✓	
Demand for goods	✓	
Site costs	✓	
Personal preferences	✓	

imity. Hoover considered there to be three economic advantages: scale economies – where the factory grows to reduce unit costs; location economies – where linked firms locate together to reduce transport costs; and urban economies – where by being together the market grows. **Linkages** are connections between firms showing the advantages of working together whilst Hotelling's model is a good example of the way in which **competition** can force firms to look carefully at their location.

2.3.5 How do regions respond to economic changes?

If individual firms grow and decline then so can regions. In this section there are two main aspects that need to be considered: what factors can be used to explain growth (and decline) and what can be done about either. Usually it's only decline that is seen as a problem and many government and EU policies have been directed towards this. It is possible to be overcome by success! As Barlow noted in his 1939 Commission Report on the Industrial Population, one of London's problems was that it was too successful – something should be done to make it less dominant.

Change is expensive in terms of labour, production capacity etc. and so there must be a great advantage to make it worthwhile (i.e. to overcome **inertia** or the resistance to change). This emphasis on change is deliberate – there is no way that any region can remain static. Several geographers have tried to find models to explain this. The **model of circular and cumulative** causation was proposed by the Swedish economist Gunnar Myrdal. His idea was that, essentially, success breeds success. In Figure 2.3.5 there are four inter-related cycles. The first is centred on wages where as the firm grew so wages would rise. These could be spent on other services etc. allowing further firms to settle there. A second cycle centres around jobs: new industries create a pool of skilled labour which incoming firms hire, thus attracting further trained migrants. The third cycle involves the linkages created by an increasing agglomeration of business while the fourth considers that as growth occurs so economies of scale reinforce it. Of course, similar ideas can be used to explain decline. Compare this growth model with Figure 2.3.6 where the same idea of cumulative causation has been developed, by Haggett, to explain cumulative decline. Myrdal was concerned about the growth in a region but he looked only at the positive side or the **backwash effect** as he called it. This would refer to the self-reinforcing nature of successful firms. Related to this is the notion that growth may be transferred to other places via **trickle-down** i.e. some aspects of growth would find their ways into the cheaper surrounding area bringing wealth to it. If this involves not just a branch factory but a spread of ideas then the

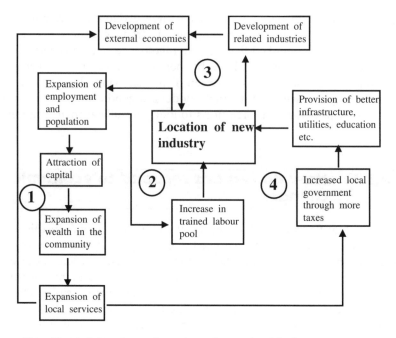

Note: The circled numbers refer to the cycles mentioned in the
text: 1 = wages; 2 = jobs; 3 = linkages; 4 = growth

Figure 2.3.5 Myrdal's cumulative causation model

diffusion effect (slow spread of novel ideas through a region) may be an
explanation.

Pred argued some places continue to grow or at least do well in
regional terms despite the changes that have gone on (e.g. Southeast
England). What happens is that as a region grows so it takes on more than
just jobs. With it comes education, finance, trade etc. He called these added
bonuses the **multiplier effect** i.e. one success would bring in other
aspects to multiply the value of the region. To some extent Pred relies on
continued growth to justify his ideas. Other writers accept that change is
part of development. Thompson has youth, maturity and old age in **industrial cycles**. A youthful situation sees firms moving into an area which has
high levels of capital and rapid innovation. By maturity, the area has
become established and dominant. Firms located in it would locate branch
plants in poorer regions because they were cheaper. In old age the advantages of the initial region would be lost as branch-plant regions would have
grown with the firms.

According to Herbert Giersch, growth would occur in the centre of a
region. Using the idea of a surface with even resources, he argued that
initially all manufacturing would be evenly distributed. However, if you
take transport costs into account then the best locations are those that
minimise transport costs. In a circle this would be the middle. Thus,

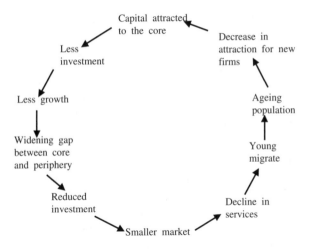

Figure 2.3.6 *Haggett's cumulative decline model*

according to Giersch, industry would go from an even distribution to a clustered one. This idea has been taken further in France and Holland where Perroux's **growth pole theory** states that leading industries would attract others (through the multiplier effect). This would also lead to the area getting more firms in a similar business. Thus the area would end up as a high-technology centre for example (such as is seen in Silicon Valley, USA or Toulouse, France). Some nations have used growth pole theory to 'plant' growth in a region to generate this kind of regional income. In another possible explanation, **export-base theory** divides the jobs in a region into two – those involved in producing goods for export and those providing services for those jobs. Large regions will produce more for export outside that region than small ones (economies of scale). That in turn will encourage people to move into the area to take advantage of the prosperity. These people need services and so the second job market will grow. So each piece of growth in the export sector will increase the service sector jobs. This can be self-reinforcing. Alternatively, a decline in export will lead to people leaving the region causing further service sector job losses etc.

In a capitalist system such as seen in the UK, profit is the key to business. To make profit the region needs to be stable in terms of regulation (national policy etc.). Sometimes, as with the business and Kondratieff cycles, there's a tendency to change from one phase to another. This in-between phase is critical. According to **regulation theory**, at this time it would be possible for a previously wealthy region to decline and a poor one to grow. This changeover time is critical: some people argue that each change brings its own geography i.e. key regions will change along with the change itself. This can be illustrated by looking at the current debate between Fordism and Neo-Fordism which relates to the nature of the production process and the implications for the firm (see Table 2.3.2 for a

Table 2.3.2 Fordism and Neo-Fordism

Factor	Fordist approach	Neo-Fordist approach
Industry	Monopolistic	Flexible, rationalist
Employment	Full employment	Casualisation
Consumption	Mass consumption	Customised consumption
Production	Assembly line	Flexible, small batch
Labour market	collectivist	Competitive
Social structure	Occupation based	Income based
Politics	Left wing	Right wing
State intervention	Liberal welfare state	Free market
Space economy	Convergent	Divergent

simple comparison between the two approaches). Perhaps the best current example of regional/national/international changes in location is **Kondratieff's long cycle theory**. By examining the dynamics of the global economy Kondratieff came up with the idea that there seemed to be waves or cycles of activity. Each cycle consisted of two parts: a growth phase and a stagnation phase. In industrial terms, the first could be seen as the growth of a new line of products; the second as a time when the market was saturated by the products. At this time, another new product could succeed in boosting the market etc. Accordingly (see Figure 2.3.7) you can trace the rise and fall of product ranges. There are two aspects worth noting here because they are important. Firstly, they tend to have a spacing about 50 years apart. Secondly, they never happen in the same place in two consecutive cycles. Some workers have traced Kondratieff's ideas back in time to find some very long waves (called logistic waves to

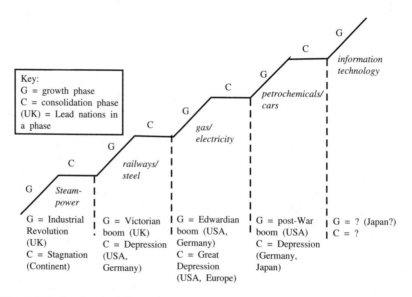

Figure 2.3.7 Kondratieff's cycles

separate them from Kondratieff's long waves). The first one started about 1050 and ended about 1450 and has been linked to the rise and subsequent decline of the feudal system. The second logistic wave went from 1450 to 1750 and is often linked with the formation of the world capitalist system we know today because it was the start of merchant trade and the development of rich/poor areas. Of course, whether you can run these waves forwards or Kondratieff's backwards is a matter of debate but it does help to demonstrate one crucial idea: development appears to occur independent of state.

> **Tip**
>
> Kondratieff's ideas are not just part of economic theory. They can be found in many aspects of geography e.g. energy, politics, settlement and are worthwhile studying in detail.

Having considered the range of ideas that can be used to explain change it is necessary to look at what can be done if the imbalance becomes too great. Most nations seem to accept that there will be some inequalities but if this becomes too great that it can impact on the national economy. For this reason, all nations (and groups like the EU) have extensive regional policies. There are two arguments about the importance of policy. One, the national-demand approach states that in a free market, firms should be allowed to rise and fall. Thus any region that didn't grow should be left to fade out. Of course, this doesn't take into account the human side of growth, neither does it have any regard for the capital tied up in the declining area. The other perspective is the theory of planned adjustment. Here the argument is that the free market does not lead to perfect decision making. In order to proceed here, you'd need to take a larger range of factors into account e.g. environment. Such a range of factors would be outside the scope of the free market and therefore government intervention would be needed.

By definition, regional policy deals with a certain part or parts of a nation. Most governments have regard for the type of region needing assistance. Aydalot's classification is a useful start in this respect. His focus on the current economic potential of the region having, at the same time, regard for its past development. For example, one of the more interesting divisions is the 'sunbelt' (or 'snowbelt' depending on where it is). This is the name given to the regions of rapid growth in new sectors e.g. high technology, and is often contrasted with the 'rustbelt' areas of declining secondary-sector manufacturing. Of course, which region is which depends to a large extent upon the context i.e. those factors which govern the development of the area. In terms of regional policy this would involve the nature of the region (under-developed, depressed or congested – too successful), the location of the region (core, semi-periphery or periphery), the nature of the industry and state of the global economy.

The next stage would be to devise a suitable policy. Of course, what's suitable depends on what the government of the day would want to do i.e. its objectives. Because this varies between governments, nations and times it's impossible to give even an overview of the regional policies that have been developed. Instead, we can see what sorts of policies are commonly available to governments. Starting with the broadest picture you can distinguish between polices which aim to boost poorer areas (**inter-regional**) compared with richer ones and those aiming to focus on a small part of a region e.g. inner cities (**intra-regional**).

By far the most common are **incentive policies** which use a number of measures to encourage firms to locate in specific areas. This can range from reduced charges to grants given to cover extra costs. These are called positive polices in contrast to the equally well-used negative policies which work by discouraging firms from locating in an area. Often, a declining region has more than one problem. It may well be that an **infrastructure** (roads and services and even industrial buildings) **policy** is needed. Some regions were able to improve their facilities and make locating there more attractive. The extensive work in the South Wales valleys is one of the best cases where UK and EU officials worked to improve the area. Alternatively, the government might want to open or takeover a firm/factory etc. This would mean that it could operate outside the usual market forces and so provide an incentive for others to locate there. One of the problems with metropolitan areas like London is that they can become victims of their own success. Costs of renting office space and the congestion in London has become too much for many firms. To reduce costs they've taken to moving large sections of their offices to cheaper, 'better' locations such as Bournemouth. With the twin advantages of reducing running costs and boosting regions **decentralisation policies** have much to recommend them.

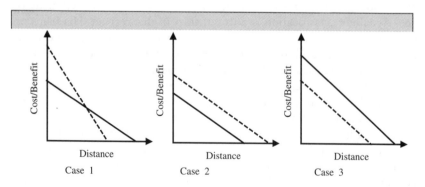

Figure 2.3.8 Externalities

2.3.6 What are the implications of economic change?

To study the effect of industrial change in social, economic and environmental terms then a point of reference is needed. This can be done by using the economist's idea of externality. An externality is a side effect. It's not part of a company's accounts but it does have an impact on the wider society. Externalities can be either positive (i.e. benefits) or negative (costs). By plotting the costs and benefits on a graph or map it's possible to see where the implications are greatest. Take Figure 2.3.8. In case 1, the cost is greater than the benefit close to the site but less farther away (e.g. manufacturing plant or airport). In case 2 the benefits are always greater than the costs while in case 3, the opposite is true.

> **Tip**
>
> When considering externalities always examine the social, economic and environmental implications remembering that externalities are both positive and negative.

Questions

1. Taking one region or sector in decline and one in growth compare and contrast the externalities.
2. Chose one business from the manufacturing and service sectors. Compare and contrast their locational factors.
3. Using a large region with which you are familiar, draw a series of maps showing the distribution of industry through time. Which of the locational models best explains what's going on? Why?
4. Write down the advantages and disadvantages of each of the locational models.
5. Choosing a region, outline the context for its development. Which are the most important factors and why?
6. Using the same region(s) you examined earlier, describe the regional policies applied to it in the last 50 years. What were the functions of the policies (use the classification we've just covered)? How well do you think they have worked? Give reasons.

2.3.7 References

Chapman, K. and Walker, D. (1987) *Industrial Location*. Blackwell.
Healey, M.J. and Ilbery, B.W. (1990) *Location and Change*. Oxford

University Press.

Wheeler, J.O. and Muller, P.O. (1986) *Economic Geography*, 2nd edition. Wiley.

2.4 Transport

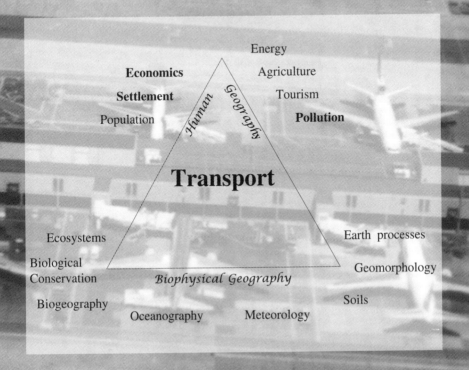

Energy

Economics   Agriculture

Settlement   *Human*   *Geography*   Tourism

Population   **Pollution**

Transport

Ecosystems   Earth processes

Biological
Conservation   Geomorphology

Biophysical Geography

Biogeography   Soils

Oceanography   Meteorology

Subject overview: transport

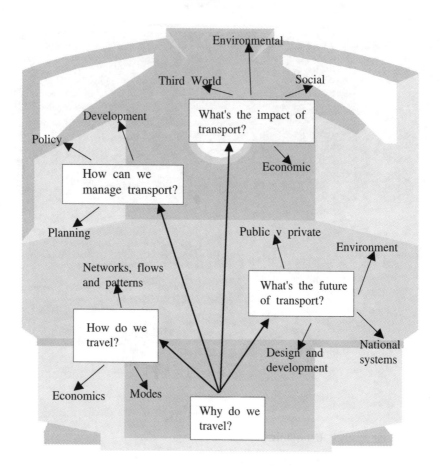

1. Transport is one of the most controversial and least studied areas of geography.
2. Travelling has a number of functions apart from the movement of goods and people. Its links with development and its use as a political weapon are just two of the alternative functions.
3. Transport is highly personal. Some writers argue that it has a great psychological value in addition to its economic one. This notion has been used to explain the lack of development in altering people's transport habits.
4. Transport models usually focus on the form of transport (mode) and the route (network). These are often explained by using classical economic theory.
5. Transport management usually involves dealing with either mode or network. Increasing difficulties with unrestrained car growth have led to schemes trying to halt this rise.
6. Transport policies tend to favour one mode. There are very few schemes which try to integrate more than one mode (Sydney, Australia, links ferries, buses, cars and trains in a combined public policy but this is unusual).
7. The impact of transport centres around environmental, social and economic issues. Some debates e.g. greenhouse effect, cover all these areas.
8. Although often seen as just an aid for development, transport in the Third World is unique. For successful implementation local factors need to be taken into account.
9. Current issues in transport centre around economics. Public v. private is a question of investment; environment is about the cost of clean-up and transport policy is seen as moving towards the most cost-effective solution.

2.4.1 Introduction

Transport is one of the big issues of the 1990s. It has been the cause of demonstrations and protests. It is not surprising that transport is so controversial: it affects all our lives. Imagine that we had to do without any form of transport except walking. How do you think our businesses would continue? At the same time, millions of people in some of the poorest regions of the world are tackling just those issues – only for real. Because our lives are bound up with movement it is important we look closely at the topic.

2.4.2 Why do we travel?

Transport is so important it doesn't have just one function but many. Thus the reasons to travel are tied up with the functions. For example, it's for the movement of goods and people. Transport forms the links that enable any form of human activity to take place. Secondly, it can be seen as an indicator of development – because of the importance that many people attach to trade transport is seen as a key factor in national growth. The opposite also applies – lack of access to transportation is a key determinant in regional poverty . Thirdly, it gives access to development. Many areas of global resources (e.g. iron ore, oil) have only been accessible by the use of transport systems. Transport can also be seen as a political weapon – transport is not neutral. Oil embargoes, where oil flows are stopped, can ruin economies. Most social groups have a way of showing status. In many parts of the developed world status is shown by personal transport. Functions, such as shops, have become set up especially to cater for the road user. Finally, transport can be seen as a provider of personal well being. One special example is transport for recreation (or perhaps transport as recreation). It has also been suggested that transport has a serious psychological value. Some writers argue that the psychology of transport is so much part of our social scene that it is almost impossible to change it.

2.4.3 How do we travel?

Certainly the way we get about (usually called the mode) is important but so is the economics of our decision and the route we take (or network to a transport geographer). The **mode** is the form of transport used. Currently we have road, rail, air and water modes (and many would also include pipelines as they are for transport, if highly specialised). This could be further subdivided. In addition, each mode would have its own set of advantages and disadvantages. Which one is chosen depends on a range of factors including two to be discussed here, economics and networks. Although mode is a question of personal choice, technology, opportunity etc., the basic economics of transport are still valid. Both Weber and von Thunen used transport costs as key elements of their explanations. They argued that the cost of transport would increase with distance but that because of economies of scale, this would not be even (Figure 2.4.1). Because the cost of providing roads is low compared with the distance for short distances, roads are cheaper. At medium distances, rail becomes cheaper because although it costs more to set up, it can carry heavier loads. Finally, there's water transport. Although expensive for short distances, it is the only economical way for long-distance bulk cargo (all of these are really just examples of economies of scale). It's possible to add a range of

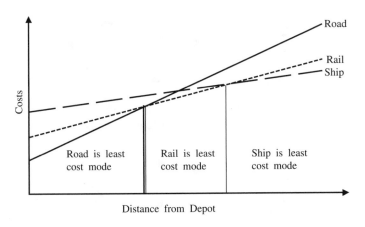

Figure 2.4.1 *Costs for transport modes v. distance*

factors which make the model more complex including the costs of transport to the hirer and provider (often for business or political reasons the true costs have been hidden with subsidies or other measures); distance rates (a graph of payment/distance would be a series of steps rather than a straight line or curve); time/direction (many companies alter their prices at certain times of the day/year to attract business at low-interest times); weight/size/bulk of goods (low weight, high value goods (e.g. jewellery) are often suitable for air freight) and handling costs.

The final part deals with networks, flows and patterns. Five factors helped shape transport networks. Firstly, there's the historical side. Much of the UK's transport system follows closely that put down by the Romans. Secondly, there's technology. There are numerous cases of new transport taking over old systems. The turnpike–canal–railway–road development in the UK has also shown how new technology can generate new transport economics (the third factor). Economics is not just cost of mode and network e.g. cars and roads, but also about problems of pricing and charging. This leads to the fourth factor, the social-political scene. Most governments have an interest in transport for taxation, social and economic reasons. How does practice relate to these factors? In the UK we have a series of major roads (not just motorways) which are termed 'strategic roads'. The idea is that during war the military will have a network that it can use easily. Railways in the UK illustrate the economic factors. During the 1950s it became increasingly clear that the rail network was far greater than needed owing to the rise of road transport. Thus a review of railways in 1963 (the Beeching Report) argued for the dramatic reduction of rural, and other, rail services. Although this has been done for almost 30 years the irony is that we now need those links to increase rural access and decrease car use.

It is common to describe network models as a series of straight lines

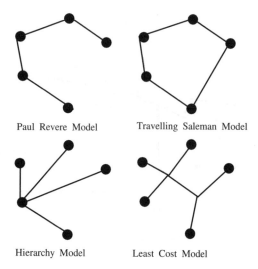

Paul Revere Model

Travelling Saleman Model

Hierarchy Model

Least Cost Model

Figure 2.4.2   Network models

(called **edges** or **links**) joined to points in the network (called **vertices** or **nodes**). As the American geographer Bunge worked out, there are a series of ways in which a set of nodes can be joined by links (Figure 2.4.2). Each one has its advantages and disadvantages. The Paul Revere model suggests a single journey where speed is important. The travelling salesman minimises journey distance back to base while the centre-orientation is good if there's only connection between centre and periphery. The circuit is the most efficient for the traveller giving minimum journeys between all nodes. The branching network minimises building costs.

Of more interest in practical terms is the way in which real networks are constructed. We use a variety of measures to compare these networks (collectively referred to as their **efficiency**). To examine these in detail, first look at Figure 2.4.3. The first of these is **connectivity** – the ease with which one can move from node to node. There are a number of measures that can be used. The beta index is calculated by dividing the total number of links by the total number of nodes. The alpha index looks at the number of circuits in the network and is calculated as: $a = (e - v + 1)/2v - 5 \times 100$ (where a is the alpha index, e = number of edges and v the number of vertices.). Next there's **centrality** – the accessibility of a place. The Shimbel number is the total number of links needed to join any place to every other one on the network. The lower the number, the greater the centrality (i.e. needs fewest links to be joined to everyone – London in the case of Figure 2.4.3). The Konig number is the route with the greatest number of links from one place to another (e.g. Plymouth to Grimsby is 5 in Figure 2.4.3). Again, the lower the number, the more central the place. Finally, there's the cyclomatic number; the number of circuits in a network. It is calculated as: $c = e - v + 1$ (where c is the cyclomatic

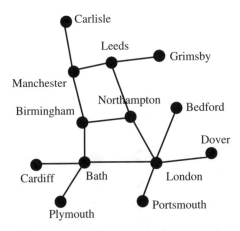

Figure 2.4.3 Communications network for England (simplified)

number, e the number of edges and v the number of vertices.). The third measure is the **density** – the number of links per unit area or total network length divided by the area. Alternatively the eta index is the total link length divided by the number of nodes. The smaller the distance the more developed the network. Fourthly one can measure the **extent** – the spread of a network. The diameter index is the number of links needed to cross a network at its widest point. Finally there's **efficiency** – the ease with which communications can be made. The detour index compares the actual distance on a road map, for example, with the straight line distance. The nearer to 1 the more direct the network.

Having examined network parameters the last part is to see how model networks have been used to explain development of an area. One of the most famous is that of Taaffe, Morill and Gould which was based on the development of transport in East Africa (Figure 2.4.4). Given that it starts at the coast this can be seen as a colonial model (⇒ mercantile model). As trade develops so links are formed inland. These develop feeders which act as foci for urban development. Finally, as the system matures, some links are upgraded as main links. Note that as the transport links mature so the settlements grow. In a similar manner, Rimmer devised a model for colonial Southeast Asia. Before contact each nation would have its own network. Early colonialism would develop coastal ports and favoured inland sites. The complexity of the situation would then develop with links being upgraded as needed. Lachene took an alternative route devising a model very similar in idea to Christaller. He assumed a uniform surface with people uniformly distributed and a series of grid-iron links. After a while some places would develop more than others (i.e. those at the cross-roads). The links between these places would be upgraded. As more places developed so would the links leading to the final pattern seen in the diagram.

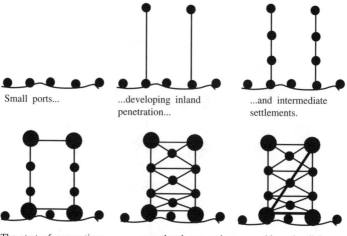

Small ports...   ...developing inland   ...and intermediate
                        penetration...   settlements.

The start of connections...   ...completed connections... ...with major links

Figure 2.4.4   *Transport development (after Taafe, Morrill and Gould)*

These three models share a common set of origins. All three are based on the physical attributes of the site even though two are based on outside influence. Gould has proposed another perspective very much in keeping with the changes in human geography in the 1960s. He put forward a behavioural model arguing that it more closely resembles the actual routes taken. Put simply, roads will be built where they are thought to offer economic advantage. In the early stages, the roads will just go out from the main settlement. As constraints and resources are met so the road pattern will diverge accordingly. Further stages of development 'open up' new areas of resources. Although one can argue that this model doesn't take into account politics it has the great advantage of linking, explicitly, transport and economics through route selection.

2.4.4   How can we manage transport?

It should be quite clear by now that the major focus of transport is economics. Some models make this quite explicit (e.g. Lachene, Rimmer). Other geographical studies e.g. economic geography have made transport differences central to their explanations (e.g. von Thunen). In this section we are concerned about how we can handle transport on a daily basis. Here the focus is not economics but community and travel. Thus although models are useful, it's planning, policy and development that make the systems workable.

We need to 'manage' transport because its growth has created a number of difficulties of which the more pressing are congestion, lack of

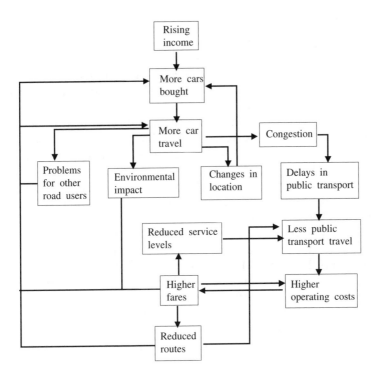

Figure 2.4.5 The vicious circle of urban transport

public transport particularly in developed nations, management of car parking, changing land use e.g. out-of-town centres, expanding demand, transport-deprived groups (30% have no access to private transport in the UK), environmental problems, difficulties for non-car users (bikes are rarely given enough space in the UK) and access (especially in rural areas with no public transport).

Such problems have a range of causes. The first of these is **economics**. Most transport problems have solutions but often the problem is money i.e. there's no balance between supply and demand e.g. the costs aren't borne by the users. The economic problem in the developing world is absolute lack of money rather than the relative level of expenditure. In many Third World cities the lack of cash effectively stops further development. In addition, there's **transport dynamics**. Transport networks might be fixed but their uses aren't. As new developments come along networks can't use them effectively. For example, transport network inertia in Bath leaves a Medieval street pattern being used by HGVs. **Individual choice** means you can't have complete freedom because you have to interact with everyone else's freedoms! One example of this has been the demand for personal transport. This comes at the expense of public transport and creates a vicious circle (Figure 2.4.5).

Now we've outlined some of the problems and causes, what of the solutions? Although there are numerous individual schemes throughout the world they can be divided into three: planning, policy and development. **Planning** usually refers to the physical arrangement of the network. Changes in economics or conditions have often meant a change in the network. Some, such as the reduction of railways in the UK in the 1960s was just a question of closure with some upgrading (more a kind of negative planning). Other ideas have met the challenge of increasing demand by structuring the network (e.g. neighbourhood concept, Radburn principle ⇒ settlement). If this type of planning tries to limit networks and therefore users by size, another example tries to separate networks. To encourage bike users, many mainland European towns have a separate cycle network reducing accidents and promoting this as an easy alternative to the car. **Policy options** tend to fall into one of two categories: flow or network. Flow policies seek to control characteristics of networks. Examples of this include vehicle weight restrictions, freight rates and driver's hours. Network policies try to create or remove parts of networks. Here, the best example would be the UK motorway programme. If regulation does not solve the problem then there are alternatives.

Network solutions tackle the problem of congestion and public transport. One example is the 'zone and collar' idea. The town is divided into a series of roughly circular zones. Between these zones there is a series of traffic lights which restrict movement (the collar). Movement around the zones is fairly easy but any attempt to get towards the centre (through the collars) meets with increasing delays. Sometimes public transport has privileges e.g. right of way at crossings, not affected by traffic lights. The effect is to make the journey into town so unpleasant that the driver will chose a public mode. Another example is the 'park and ride' scheme – a mixture of network and modal control. The idea is to drive to an out-of-town car park and then get a bus into the centre. **Modal schemes** tend to look at the types of transport needed. Such schemes are commonly referred to as unconventional modes. Dial-a-ride schemes are spreading as their attraction to less mobile families becomes more problematic. Postbuses are popular in rural Wales and Scotland.

2.4.5   What's the impact of transport?

Most transport geographers would recognise three main impacts: environmental, social and economic. In addition, we should study the impact of transport in the Third World. This is often neglected and yet there are some valuable lessons to be learned.

The **environmental impact** of transport has been grabbing the headlines for years: there's no doubt that traffic has an impact in a number of key areas. The first of these is **energy consumption**. An increase in the demand for mobility coupled with the rise in personal transport has meant that energy supply is a key issue in transport. Not only is the total amount of energy needed under discussion there is also the question of energy efficiency. To meet this challenge most manufacturers have tried to make more efficient engines. **Pollution** is another major impact. Transport is seen as a key polluter in many countries. Carbon and nitrogen oxides, both byproducts of combustion are adding to global warming as both are part of the so-called greenhouse gases. Other gases, hydrocarbons, lead, aromatic organic chemicals combine to give a cocktail of pollutants which, in some of the world's largest cities, exceed the statutory limits. Although there have been attempts to reduce this by the use of unleaded fuel and exhaust controls there is still a great deal to do. Not all pollution is aerial. **Noise** and **vibration** is another issue. Noise, particularly around airports had led to a number of measures to combat it including changing engine design, altering flight patterns and soundproofing houses. Whilst this noise is disturbing, vibration can be a greater problem. If the frequency is close to that of a building then the building can develop cracks and subsidence. In some historic areas such as the centre of Amsterdam and Rome, traffic vibration has so threatened the area that vehicles are banned from certain places. **Land-take** is the amount of land needed for a project. This can be divided into primary and secondary land-take i.e. needed for the project directly (road and runway) or indirectly (quarry for materials, areas polluted by traffic). If this also includes areas 'blighted' by transport (i.e. seriously affected such as around airports) then significant areas are being used. Fifthly, there's the **visual impact** of transport. Visual impact is controversial – many people protested when railway bridges were built across moorland; today, people protest if they are threatened with demolition.

The social impact of transport is less often heard but is just as important to study. For example, there is often the **alteration of social conditions**. The introduction of mass travel in the nineteenth century. meant that for the first time working people could enjoy some recreation. The growth of the seaside resort was a result of that policy. Blackpool, described as a tiny village with a dozen boats in the 1830s, was just one of many places whose economy was transformed. The extension of public transport in the 1930s led to the idea of commuter settlements and the notion of travel to work. For people moving outside London and travelling in by train the whole idea of transport was as a lifestyle rather than a means of getting from a to b occasionally. Such changes are rarely commented about in the literature but are fundamental to our understanding of transport issues. Secondly, there is the notion of a **mobility gap**. Many people take transport for granted in the UK and the developed world. However, studies have shown that there is a remarkably similar percentage – about 30% – that do not have access to personal transport. Given the reduction in public provision in both the UK and the USA we have to wonder what such disadvantaged groups will do. In the developing world this situation is even more difficult with poor provision often made worse with poor infrastructure. Further, many transport projects are designed by Western firms which may not always be focused on the wider social issues or understand the culture they're working in. So far, this aspect has considered those with no access such as the poor. But there are other groups: elderly, disabled, children and women. Each has their own requirements. Most transport is run for those with complete mobility and certain social needs i.e. work. Those without such demands may find provision less than good. **Accidents** are one of the negative sides of transport. Every mode has some risk attached to it. Rail might be the safest and road the least safe but it does depend on a range of factors. Currently, there is considerable effort in designing transport systems which provide maximum safety. Transport also impacts on **social relationships**. Early studies of the geography of perception noted that poorer people with little access to transport had limited geographical boundaries. Often in the past, road schemes have been designed without regard to social conditions. Many communities were divided by major road schemes in the 1950/60s. Today, with an increase in the interest in social geography, social relations have been studied. Now it would be unlikely that a neighbourhood would be divided. Lastly, there is the issue of health. One of the big pollution stories of the 1970s was the argument against lead in petrol. A toxic heavy metal, lead was linked to problems in brain development. Studies showed that children living near major roads were far more likely to have lead levels in their blood which could suppress brain function i.e. they would appear less able because of the damage lead could do. Other studies linked lead levels to violence. After a considerable battle, petrol companies removed lead from petrol.

The third impact to study is the **economic impact** of transport. There have been numerous studies that have linked prosperity to transport links. Swindon was a key railway town. The loss of the railway yard (once the biggest in Europe) was a major problem. Other railway towns such as Crewe and York have faced similar difficulties. The M25 was designed to help the economy of the London area and to reduce congestion in the centre of London (despite the problems faced since its opening). Although road transport is a major issue in economic development there are other examples. The debates over the expansion of Heathrow and Gatwick airports are worth studying in a regional context. Such places are major employers as well as transport links: their impact on the local economy is considerable. Although according to economic theory the cost of houses should go down near an airport it actually increases near to Heathrow because of the job opportunities!

Finally, the impact of transport in the Third World can be assessed. Although the impacts we've been looking at in this section also apply to Third World towns and cities there are some writers who believe that there should be a separate study (much like urban models). They base their ideas on the following:

1 Third World nations have a greater diversity of modes. In many, non-motorised modes (walking, cycling) are dominant. This diversity is not just between urban and rural but between nations. There is no such thing as a 'typical' or 'average' transport situation.

2 Most theories of development follow the so-called modernisation theory. Here the basic idea is that technology is politically neutral and leads to an increase in economic wealth. Neither statement can be fully proven.

3 Most ideas of transport development came from universities and developers in the North. These solutions might be fine in their specific nations but do not pay regard to the completely different setting in the South. Physical, social and economic environments are often completely different. Thus whereas the vast majority of the transport schemes are North-funded there is little evidence that they take local conditions into account.

Even from this brief look at Third World transport it is obvious that there is far more to do than just translate developed-nation ideas.

2.4.6 What's the future of transport?

There are really very few avenues to explore: most deal with extensions of current design or ideas. For example, take the case of **public v private** investment. Among the developed nations there is some agreement that public investment should be reduced (for the EU this is part of the transport policy). This has gone furthest in the UK where the privatisation of road and rail systems has led to some interesting outcomes. The idea of efficiency has not always been achieved and the costs have been greater than anticipated. Questions about funding lie at the heart of transport planning and so should be seen as crucial. Private investment should be 'efficient' in terms of conventional economics but against this is the idea of transport for a nation and the reduction of pollution by using more fuel-efficient means.

A second way forward concerns the problems of the environment. The debate about lead in petrol was finally won but there are still numerous issues. Many cities in both developed and developing nations face serious congestion. Add to this the high levels of pollution and many cities are faced with a brown haze from photochemical reactions. To reduce this problem there have been calls to ban traffic, change to electric vehicles or any of a range of ideas. There is a clear issue to reduce the negative environmental impact but how this is achieved is less certain.

One of the difficulties seen with UK road policy was its fragmentation i.e. there was no coherent overview. Many nations do have a public system with many modes integrated. This is seen as a key way of getting greater use: integrated timetables and single multi-mode tickets reduce the problems for commuters and so boost potential usage. Other schemes, backed by national computer systems, have tried to improve efficiency. For example, most freight vehicles are travelling back to base empty. In theory, if some of these could carry suitable cargo, then the number of trips could be reduced.

Despite all the advances in technology few have been aimed at reducing fuel consumption. Today, with higher fuel prices this is now a concern. There are 'lean burn' engines, better jets and aerodynamic shaping. This doesn't tackle the issue of how the remaining petrol supplies are shared out but it does allow for more efficient usage.

___ **Questions** _____

1. Outline the strengths and weaknesses of these various measures of network analysis. Work out the results using Figure 2.4.3 and compare that with a model of your local area. Explain any similarities or differences.
2. Outline the advantages and disadvantages of the key transport modes in your area.
3. Using your local area, gather information about the transport management policies of the local council. Why were they put in place? Have they solved problems or created more?
4. Does London need another airport? Would a fifth terminal help at Heathrow? Could regional airports be developed to cope?
5. Does Britain have an integrated transport policy?

2.4.7 References

Hoyle, B.S. and Knowles, R.D. (1992) *Modern Transport Geography*. Belhaven Press.

Simon, D. (1996) *Transport and Development in the Third World*. Routledge.

Tolley, R. and Turton, B. (1995) *Transport Systems, Policy and Planning*. Longman.

Waugh, D. (1995) *Geography: an Integrated Approach*, 2nd edition. Nelson.

Wynne-Hammond, C. (1986) *Elements of Human Geography*, 2nd edition. George Allen and Unwin.

2.5 Energy

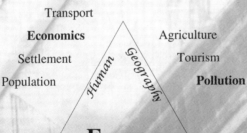

Transport

Economics   Agriculture

Settlement   Tourism

Population   **Pollution**

Human Geography

Energy

Ecosystems   **Earth processes**

Biological
Conservation   **Geomorphology**

Biogeography   Soils

Biophysical Geography

Oceanography   **Meteorology**

Subject overview: energy

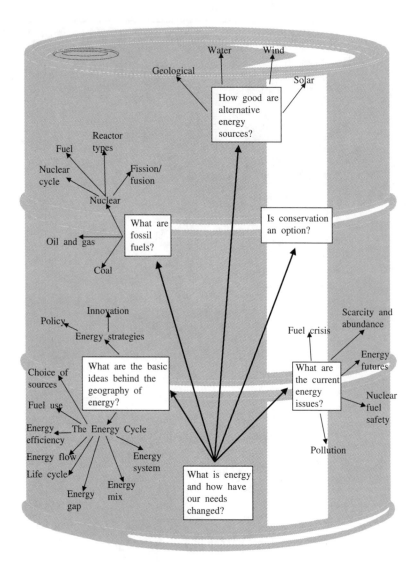

1. Energy supply and usage is one of the most crucial studies of modern times.
2. Energy resources can be classified in a number of ways. From a scientific perspective we get potential and kinetic energy; from geography we can divide energy resources into finite or renewable.
3. Energy is never actually 'lost' in the system but energy resources do degrade from 'concentrated' forms like petrol to the most 'dilute' form, heat.
4. As societies become more complex so our energy demands increase.
5. Energy industries develop in a set cycle from discovery and exploration to depletion. We can see different sources follow this pattern throughout history.
6. The length of time a resource lasts depends on a range of factors – technology, legislation, cost, society, resources and location.
7. In any analysis of energy, two key areas are examined: the energy mix (how it's made up) and the energy gap (a shortfall in supply). If the analysis were national then the security of supply would also be important.
8. Fuels (finite and renewable) can be compared in terms of formation, usage and implications. For energy conservation a major aspect is payback time – the time needed to repay the cost.

2.5.1 Introduction

The geography of energy can mean different things to different people. To the motorist it's the cost of petrol and the queue at the filling station. To many a person in the developing world it means the daily chore of finding enough fuel to cook the evening meal. In this chapter the whole issue of energy comes under scrutiny from basic concept to environmental impact.

2.5.2 What is energy and how have our needs for it changed?

Energy can be divided into two sorts – kinetic and potential. **Kinetic energy** is the energy that a moving object has. If you try to stop a ball rolling you have to use energy to do so. **Potential energy** is that stored in an object because of its chemistry or position. Water flows over the edge of the waterfall into the stream below. What happens in energy terms? The water at the top has potential energy (it's above the stream and can fall

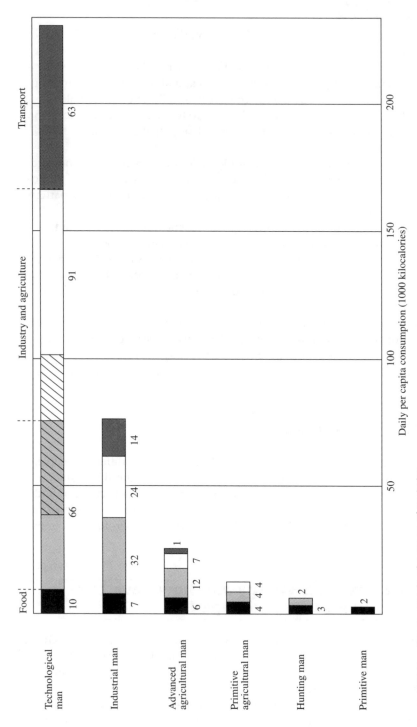

Figure 2.5.1 Energy needs of a range of societies

under the effect of gravity). As it goes over the top it loses potential energy which is converted into kinetic energy. When it reaches the bottom it has lost its potential energy and the kinetic energy is used when it meets the stream water. The ways in which energy changes between one form and another is explained by the **laws of thermodynamics**. Energy can be neither created nor destroyed (suggesting a finite amount). As it changes form it 'degrades' i.e. goes from a 'concentrated' to a 'dilute' form. Heat is the most dilute form. From these laws there comes the idea that at each energy transfer, some is 'lost' (called **conversion losses**). For example, an electricity generator is about 33% efficient (which means that out of 100 units of fuel it produces 33 units of electricity).

Human energy needs have changed through time. Figure 2.5.1 shows the energy needs of a range of societies (implying a development from the hunter/gatherer to industrial society). In the earliest societies our needs were purely for food. A more 'advanced' society would need more energy for fire. Continue this through the different societal groups and it is possible to see an increase in need for energy (adding transport, industry etc.). What it shows is that energy needs change dramatically with societal group and that energy needs are often unrelated to the individual but refer to the society as a whole.

2.5.3 What are the basic ideas behind the geography of energy?

There are numerous sources of energy. To make comparisons easily there is a need for some basic concepts of which the most important are: classification, the energy cycle and energy strategies. Figure 2.5.2 shows one way of classifying energy resources. The main division is into those resources that have a fixed **lifespan** (called **finite** or **non-renewable** resources) and those where, for all practical purposes, there is no limit (**non-finite** or **renewable** resources). These divisions are important not just for the amount of resources but in the way we use them, their efficiencies and environmental impacts.

The **energy cycle** is the process of usage of an energy resource from discovery to depletion (i.e. no more economically available supplies). Given that this is an economic system we can represent it as an 'energy system' made up of a number of variables (see Figure 2.5.3). At the heart of the system is the **energy supply chain**. Around this core there are a number of other components: finance, organisation, spatial distribution, and the environmental and social dimensions of energy supply. This should be seen as a template to be used when comparing and contrasting energy forms or nations for example.

The production of an energy-cycle diagram (including quantities,

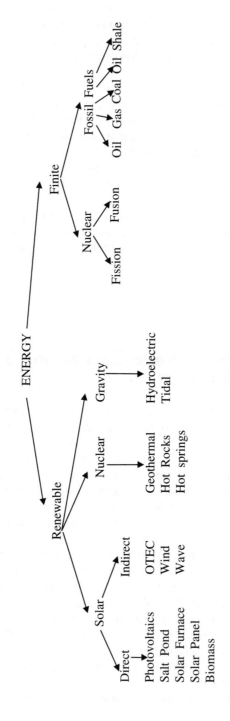

Figure 2.5.2 Classification of energy resources

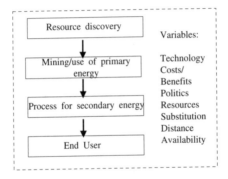

Figure 2.5.3 An energy system

dates etc.) relies on a range of information most of which changes through time. Variables which alter the energy picture include technology (better equipment); legislation (to influence the way in which energy is used); cost-benefit ratios (to check viability); social values (what society expects from its energy supply system); resources (how much is left) and location and population density (where resources are and how their distribution is going to be handled). So far, the energy cycle only refers to one resource. In reality, numerous sources are available and the availability of these changes through time. The range and quantities of the energy resources used by a society is called its **energy mix**. Figure 2.5.4 gives one example of an energy-mix diagram. Note that if there are projections into the future there might be a difference between the amount available and that demanded. The resultant **energy gap** is crucial because it implies a short-fall which must be met by a limited number of strategies. There is a third way of examining energy resources: to see how they are used in a nation at one particular time. **Energy-flow diagrams** are one way of displaying this information (Figure 2.5.5). They show not only which forms of energy go where but can also highlight other features of the system such as conversion losses overall (60.2% in the case of Figure 2.5.5) and, by comparing yearly figures with GDP, the **energy efficiency** of the system.

Energy strategies examine ways in which we can use our resources more efficiently: broadly speaking, policy and innovation. The former is concerned with the ways in which government ideas can impact upon our use of energy; the latter examines the changes that technological development can have. There have been numerous attempts to control energy usage; some have been more successful than others. If we wish to make sense of any policy moves we need to remember that policy does not take place in a vacuum. Indeed any policy must work within a number of constraints: political context; perception of the issues (a major problem with many of the more complex environmental cases); policy making; policy content and policy monitoring. How have different nations responded to their own policy needs? Look at Table 2.5.1. For the former centrally planned economies like the Soviet Union, the key was to increase

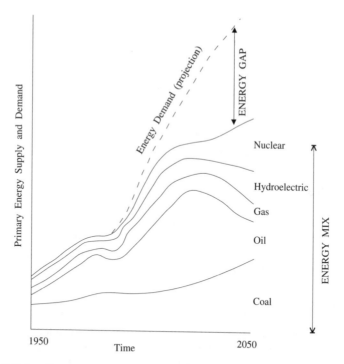

Figure 2.5.4 Energy mix and the energy gap

production and keep internal self-sufficiency. For the USA the key aims were stability and security (i.e. a given level of supply for a given time). Here, the location of the supply was less important. For Japan, the aim was also to get a stable supply but, unlike resource-rich USA, there was the need to reduce the amount of oil needed (hence their larger nuclear power programme). In contrast, OPEC (a group of producer-nations, the Organisation of Petroleum Exporting Countries) were chiefly concerned with getting the maximum benefits from their oil wealth and keeping their income flowing. The most important aspect to remember here is that strategy depends upon perspective.

2.5.4   What are fossil fuels?

This term is used because most of the finite fuels have their origins as 'fossils' i.e. coal is made up of plant remains and oil/gas are animal remains. The first of the fossil fuels is **coal**. It was one of the earliest of the modern fuels having some use from Medieval times onwards but only really coming into its own during the Industrial Revolution. Coal formation is fairly straightforward. Initially, plants die and start to decompose in

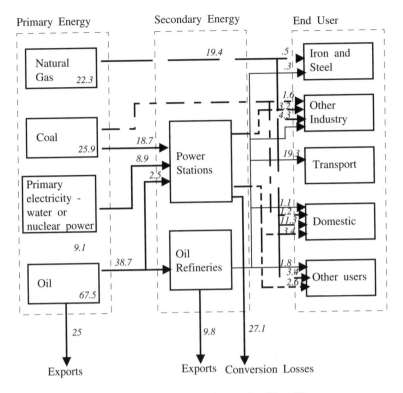

Primary Energy Secondary Energy End User

Natural Gas 22.3		19.4	.5 Iron and .3 Steel
Coal 25.9	Power Stations		1.6 Other 3.2 Industry 4.3
Primary electricity - water or nuclear power 9.1	18.7 8.9 2.5		19.3 Transport
Oil 67.5	Oil Refineries	38.7	1.1 1.2 Domestic 11.3 3.4
			1.8 3.4 Other users 2.6

Exports 25   Exports 9.8   Conversion Losses 27.1

Note: Only major flows shown. Figures in thousand million Therms

Figure 2.5.5   UK energy flows 1992

an oxygen-free (anaerobic) environment. The classic place for this to happen (such as during the coal-forming Carboniferous period in the UK about 350–200 million years ago but also elsewhere in another coal period from 150–50 million years ago) is in a swamp. This process of gradual breakdown is called **decomposition**. There follows the build-up of decomposing matter over millions of years. At some stage, the conditions will change and this rotting vegetation will be covered with silt and clay. At this stage another process, **coalification**, can begin. The idea here is to increase the amount of carbon relative to oxygen. The different lengths of time and pressure for this process gives rise to a **rank sequence**: peat to brown coal, steam coal, bitumenous coal and, best of all, anthracite.

Exploitation of this resource would depend upon its location. Early use of coal came from exposed seams in river banks or on beaches. Later, the seams were tunnelled into (drift mining) whilst with improving techniques in exploration (geological mapping and seismic survey), shaft mining could reach deeper seams. Today, it's often viable to strip off covering rocks and soil (overburden) to surface mine coal seams. Whether modern mines use shaft or surface mining there are implications:

Table 2.5.1   Energies Strategies

Nation/Group	Aims	Strategies
USA	1. Supply stability 2. Supply security 3. Supply strength	1. Deregulate industry 2. Maintain reserves and encourage exploration 3. Diversify mix
Russia	1. Increase production 2. Increase efficiency 3. Increase exports	1. Plan increases 2. Decrease consumption/ substitute other sources 3. Maintain exports
Japan	1. Secure supply 2. Decrease oil dependence 3. Increase efficiency	1. Diversify supply 2. Increase mix 3. Reduce consumption/ replace equipment
OPEC	1. Maximize benefits 2. Maintain income 3. Modernize economies 4. Assist developing economies	1. Concerted action 2. Control price 3. Develop infrastructure 4. International funds
OECD (International Energy Agency)	1. Coordinate developments 2. Increase efficiency 3. Environmental policy 4. Decrease vulnerability 5. Increase local supply	1 Annual meetings 2. Comparative studies 3. Establish committees 4. Encourage substitution 5. Maintain free market

- Geology – shaft mines need props to support the weight of rocks overhead leading to a small percentage of the coal mined (typically 25–50%). Alternatively the seams might be far too thin to mine economically or might possess geological faults (or even buried rivers causing flooding).
- Economics – coal is a finite resource and as such there are economic implications of using it. It soon became the main fuel of the Industrial Revolution replacing the earlier and less efficient water power. Today, with the rise of electricity and gas, there's less demand for coal (even at power stations) which means considerable implications for coal-mining areas.
- Environment – much of the difficulty with coal has been the environmental implications that have arisen during the twentieth century. Much of this is related to air pollution e.g. acid rain, global warming ($\Rightarrow$ pollution) although there are other pollution effects: health (e.g. London smog 1952); dust pollution and radioactivity (actually more than most nuclear stations).

Petroleum is the general name for a range of liquids and gases. Like coal they must be formed in anaerobic conditions but unlike coal these are best found in deeper sea environments. Initially, this means the accumulation of dead animal remains in deep ocean basins with anaerobic conditions and sediments (usually sand) deposited as well. In this early phase there would be some initial breakdown of the decaying matter forming methane and kerogen. In the second stage there is the burial of the kerogen usually with clay. This would increase both pressure and temperature allowing further chemical processes (distillation) to take place. The final product would eventually be a mixture of crude oil and methane gas. If geological conditions allowed accumulation of the oil and gas (e.g. a fold) then sufficient stock would be available to exploit. Exploitation would, like coal, depend on the location of the resource. Natural seepages of oil were known by Amerindians for example but were only fully exploited later. The problem is to find reserves buried deep beneath the Earth's surface. Early work used geology maps but with the discovery of aircraft-based magnetometers (very sensitive magnetism detectors) it was possible to detect the variations caused by less dense oil fields and salt plugs. From such large-scale detection came the seismic survey and finally the drilling. Depending on the location, the final stage would be to pipe/ship the crude oil to a refinery.

What of the implications? As with coal these have changed over time but today the more common arguments are:

- Geology – increasingly the fields are in difficult locations. The North Sea fields first opened in the 1970s were a key source but an extremely hostile environment. The Gulf of Mexico is proving no easier and with new fields in the Far East around Sabah and NW Australia oil platform technology is being stretched.
- Economics – have been the key to understanding oil usage. The cost of exploration and recovery increases as the area becomes more difficult to work in. Early Texas fields would be very simple affairs but modern deep sea platforms cost millions of pounds. Also, oil has been used as a political/economic weapon. When Middle Eastern oil was first discovered it was controlled by the oil companies. In the early 1970s the oil nations formed OPEC, the Organisation of Petroleum Exporting Countries – an oil cartel could, for the first time, control supplies. In 1973 they raised the price of oil three-fold which had disastrous implications for both developed and developing nations. Such a move occurred again in 1979. Because of the importance of oil to the modern economy there were some very serious political repercussions.
- Environment – now the key factor in discussing oil and gas. Being a hydrocarbon most of what can be linked to coal is also true of oil and gas. In addition, there's the problem of mass oil loss through tanker accidents ($\Rightarrow$ pollution).

Nuclear energy is one of the most controversial energy sources. Before starting to look at the nuclear process there are two things to note. Firstly, all things are radioactive to some extent (see Table 2.5.2). The second point is that all radioactive substances (those isotopes of lighter elements

Table 2.5.2 Background Radiation Values

Source	Amount (mrem)
Human body	2–20
Cosmic rays	25
Ground gamma rays	35
Radon in air	117
Medical sources	30
Fallout	1
Industry	2
Nuclear waste	0.2
Nuclear reactors	0.01

Note: Approximately 85% of radiation we receive is "natural" i.e. from the environment

and all heavier elements such as uranium) will decay. Because it's a random process we don't know when any one atom will actually change but we do know that for each radioactive substance the time taken for it to decay to half its original mass is a constant for that substance. We call this the **half life** so 100 g. of uranium will decay to half its original mass in one half life (it'll go from a half to a quarter in the same time and so on). Some half lives can be millions of years giving problems in waste disposal. However, those with very short half lives also give problems because the faster they decay the more damage they can cause. The nuclear reaction is used to produce electricity but it requires a great deal of engineering to make it safe. The aim is to create is a continuous reaction (called a **chain reaction**) where the next decay is made possible by the last. If this is going to happen enough material is needed to keep it going (the **critical mass**). Once this can be achieved the aim is to keep it going whilst still being able to control it safely. Figure 2.5.6 shows a basic reactor type. There are many variations to this depending on the fuel type, coolant type and operating pressure. What are the implications of nuclear power? Some of the key ones include:

- Geology – concerned with the obtaining of uranium and its disposal after the fuel cycle has been completed. Often the production of uranium has considerable health implications. Once it has been used there is a need to dispose of it. Sometimes it is possible to recycle it in a re-processing plant and other times it must be stored. Both these processes have problems. Re-processing can lead to pollution. Burial can be a solution but the storage place must be safe i.e. watertight etc. and deep enough to contain the radiation. One method, deep sea storage, was tried. It put radioactive waste in concrete which should have lasted 600 years. Sadly, after only about 20 years it started to break up!
- Economics – it's best to consider these in the context of the nuclear fuel cycle (Figure 2.5.7). Part of the cycle is the mining of ore, its enrich-

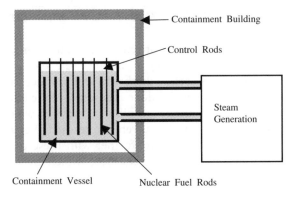

Figure 2.5.6 Basic reactor type

ment (making the uranium concentrated enough to sustain a reaction) and its usage at a power station. Up to this point the economics are much the same as other electricity generators. However, there are the disposal costs to consider. This is material you don't dump but look after for maybe hundreds of years. The cost of doing this is considerable. Some calculate that it costs more energy to use and store nuclear material than it produces in electricity.

- Environment – this is the area that gets the most public concern. Accidents such as Chernobyl do release pollution into the atmosphere. In addition, most nuclear power stations produce a small amount of pollution through the cooling water. Generally, such amounts are within limits set by international agencies.

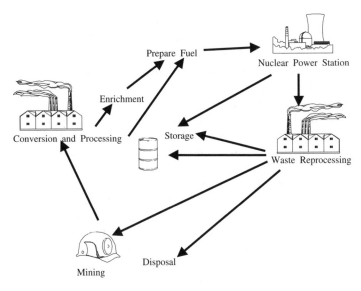

Figure 2.5.7 Uranium fuel cycle

2.5.5   How good are alternative sources?

Although we call them alternative (or renewable) sources of energy, which implies that they are modern, they're amongst the oldest examples of energy use. Solar power was used to dry bricks and produce salt in the time of the Egyptian empire. What we've done is to improve the efficiency of the scheme. Today, there are a number of different ways in which renewable energy can be generated. Given the current interest in the environment it's worth asking why we haven't got any further with these energy sources. The answer to that lies in examining six aspects which are common to all alternative sources:

1 **Amount** – the total may appear infinite but the usable portion is finite. Put it this way, if you site too many wind generators together you'll eventually run out of wind to turn them! Of the vast amount of energy that reaches the Earth we can only tap into a portion;

2 **Location** – unlike finite energy sources which have a certain amount of locational movement there's no such luxury with renewable energy. If you want wave power then you need to go to the coastal areas where there's sufficient energy. Solar power is no good in the Arctic in the longer term. Similarly, you must take whichever method is available in the area, so some methods are just not going to be available. Another locational aspect is the more 'portable' nature of the energy. Conventional power stations can be centralised by bringing resources to them to permit economies of scale. The more 'dilute' alternative energy systems must be where the source is which means that they are often decentralised.

3 **Duration** – it's no good having an excellent source if it's very short lived. Some sources like tidal power have two or four peaks of generation and, of course, solar power is not possible at night! This means that there's no overall control of the flow of energy. All this comes down to...

4 **Storage** – once you've heated up your water tank, for example, any further energy would be unused. Renewable sources are notorious for being very short term. The aim is to get hold of this energy and keep it until it's possible to use it. There have been many attempts from the rechargeable battery to growing plants and harvesting them for power generation (biomass). One of the more promising ideas is the creation of hydrogen through energy usage e.g. break down water to oxygen and hydrogen and store the hydrogen.

5 **Efficiency** – the greatest drawback for all renewable energy resources is their efficiency. A gas fire may be up to 70% efficient. Even the poorest power station can manage at least 28%. Compare this with the 2–10% range for the renewable resources. Thus not only do we need more energy collectors but they must cover a far wider area. In terms of efficiency we need also to consider...

6 **Economics** or **Pay-back** – It's no good having a superb form of pollu-

tion-free energy if the resulting technology needed will be far too expensive. One of the arguments against alternative sources has been the cost per unit of energy compared with conventional sources. If the energy being discussed is electricity then the price/energy figure becomes pence/kilowatt hour. In the debate over nuclear power in the UK in the 1980s the 'target' figure was quoted at 3.5p/kWh. If the alternative source could match or better this figure then it would, in theory, be a 'better' option. The problem was that it depended on what you included/excluded in your calculations. The nuclear power stations didn't include the cost of electricity lines, the alternatives had to (which reduced their benefits!). In assessing any energy scheme a key question to ask is the cost or, given the life of the scheme, the pay-back time (the time taken to re-pay the cost, financial or energy, taken to build it).

Once some of the problems of using alternative sources are appreciated then it is possible to outline some of the most important examples.

Solar energy has been spoken of as one of the great energy options. The technology is available, well known and, usually, cost-effective. We tend to divide solar schemes into two: solar heating and photovoltaics. Solar heating uses the sun's energy to heat an object. Ideally, this object will radiate energy to the place wanted and/or store it for later. Examples include the Trombe wall where the sun heats up a large slab of concrete which then radiates energy into a building when it's cooler, solar buildings, solar water collector, solar furnace and the salt pond (virtually unknown in the UK but which could, in the right place, use the simple thermal principle of dense (salt) water retaining more energy to produce heat 'layer' in a pond for later use).

   The idea of photovoltaics i.e. electricity from light has been used for years but rarely in any large-scale way. The camera light meter is one example, the rotating shop-window display is another! Whilst the concept is simple, the economics are not. Until recently, the cost of producing enough photovoltaic panel was just too great. Now that we can produce some panels cheaply enough there is the possibility that it can be used. Solar-powered telephones in remote areas and solar-powered fridges are just two current applications.

Wind power is primarily concerned with some form of windmill. The siting of a wind generator is crucial for maximum efficiency. This means that coastal and exposed areas are going to be prime possibilities. Like solar power, there is a limit to the amount of energy that can be used. In this case it's 59%, the theoretical limit set by Betz (and hence called the **Betz limit**). Attempts to exceed this figure would run up against problems of the air trying to get past the blades. Apart from that the main limit is the wind speed. You'll need a certain amount to start the equipment generating (the cut-in speed) and beyond another limit (the cut-out speed) beyond which the equipment is damaged. Between those you get a steadily increasing electrical output. Recently, although they were said to be completely 'environment- friendly' there have been concerns raised about their noise and the visual impact of the larger 'wind farms'

Water power works by exploiting the mass of water and the action of gravity. Any body of moving water possesses kinetic energy and so a turbine or wheel placed in the way will be able to use part of that energy to produce power. Water power can be divided into three: hydroelectric, tidal and wave. All use the same principle but in slightly different ways. The hydroelectric scheme uses the movement of freshwater past a point. Simple schemes such as watermills have been outstripped by modern water turbines such as those associated with large dams or rivers (such as the Niagara Falls). Tidal power uses the twice-daily movement of water. Early schemes merely used the to-and-fro movement passing a point to produce a simple tidal power mill. Modern schemes allow water to flow in freely, trap it and then release it when the height between trapped and 'free' water is enough to generate electricity. Wave power has a slightly different approach. The aim in all schemes is to use the movement of the waves to move two coils in opposite directions so that electricity can be produced.

There is also one scheme which looks great in design but has yet to be seen as viable.

Biomass uses plant and animal material to generate heat and power. At its simplest this is a wood-burning stove or even a simple fireplace. At its most complex it's using the energy gained from photosynthesis to produce alcohol to run cars. By dividing this list into wet and dry processes we can see how the process works. Most dry processes involve the direct burning of materials. This is the most common use seen typically in many rural areas of the developing world. At higher temperatures it's possible to produce gas. Wet processes can be equally common. Aerobic digestion is the same process as some sewage treatments only in this case, you can gain methane gas to burn. Such a technique removes numerous problems associated with sewage treatment and gives a usable fuel. For this reason it can be seen on many scales from the city sewage works to the individual household in the developing world.

Finally, there are the **geological schemes**. This involves tapping the heat reserves of the Earth's crust. Schemes using this principle are among the best modern ideas, having been in service since the 1900s in Italy. Natural hot springs and geysers produce steam which then drives a turbine generator. Apart from the difficulties of the chemical-laden water corroding the pipework there are few problems. Using this basic idea but creating your own steam has been behind the 'hot rocks' ideas being tested in Cornwall and Southampton in the UK. If you drill two holes near a naturally hot rock (like granite for example) and join then through a series of fissures created by an underground explosion, water pumped down one should come up as steam in the other (and so through a turbine).

2.5.6   Is conservation an option?

Energy conservation involves using less energy to do the same work. Its potential is considerable. It's one of the most active areas of practical energy work at present and numerous schemes can be seen all over the world. There's little doubt that it could save up to 10% of our energy requirements with a pay-back time of between 2 and 25 years depending on what is done. There are so many schemes that it is impossible to describe even a fraction: all we can do here is to outline some of the more common ones based on sectors of end-use:

1 **Domestic sector.** The average house could lose 25% energy through the roof, 10% through the windows, 35% through the walls and 30% through floors and draughts. By using, respectively, loft insulation, double (or even triple) glazing, cavity wall insulation and floor and door linings this could be dramatically reduced. Some houses in a project in Milton Keynes in the UK used so little external energy that they were virtually self-sufficient. Alternatively, conservation could be achieved with better design and orientation of housing. By being able to use available sunlight energy costs are lowered. In developing nations, the development of a simple wood stove has created a considerable increase in stove efficiency (important in areas of fuelwood scarcity). Finally, a heat exchanger (like a fridge but in reverse, could use outside temperature to heat a house.

2 **Generation sector.** Electricity generation is about 30% efficient. In other words it's 70% inefficient! Some schemes actively use this to promote better energy usage. The combined heat and power scheme (CHP) sites a generator near urban areas. The hot water produced during generation is pumped through the buildings to heat them. This saves on both the heating costs of the buildings and the cooling costs of the power station.

3 **Commercial sector.** Many ideas here are common to housing. There's an increasing use of insulation and energy-saving measures to reduce costs in larger buildings.

2.5.7   What are the current energy issues?

Just about every matter is a current issue because of the controversy energy usage generates. Some of the bigger current issues are:

1 **Energy futures**. This is important because it examines where we're going to get our future energy mix. It might just seem like a question of doing a bit more research but there are several questions that need to be asked:

   i   what is the cost-barrier to adopting a new energy source?
   ii what is the social process of adoption (i.e. innovation)?
   iii is it feasible for mass production?
   iv have we solved the technological problems?
   v   what is the energy market like – is it viable?
   vi what are the environmental impacts?
   vii are there any public conflicts/interest groups?
   viii does it work in principle (for longer-term ideas)?
   ix is there enough research?
   x   do we want to follow this path?

2 **Timing**. This is linked closely to (1), above. Not all the strategies we'd like to use are available now. When the technologies are available (and assuming they work) is a matter of debate. What many people are asking now is what happens between now and then. How do we deal with any energy gap? One measure is to use taxation.

3 Nuclear power **safety**. After the accidents at Three Mile Island in the USA and Chernobyl in the Ukraine there were numerous fears about nuclear safety. This highly complex issue is difficult to get into because of the technical points involved. Even allowing for this, it does attract a great deal of attention probably because of fears about radiation.

4 The **fuel supply**. In 1973 OPEC raised the price of oil overnight. Despite the fact that it was their oil and they could raise the price to anything the free market could afford the repercussions were consider-able. Immediate talk of everything from complete collapse to war were discussed but eventually it all calmed down. It did highlight two things though: **security of supply** and **energy demand**. Security of supply drove many countries to explore nuclear options as being 'homegrown' and free from outside influence. At the same time it allowed many in the developed world to ignore one of the true problems – the real energy crisis in the developing world – the shortage of fuelwood. Scarcity and abundance might seem like absolute terms but in fact they're relative. For example, in 1973 it was thought there was a scarcity of oil (it was actually about a 5% drop in supply!). When the North Sea oil production started it was given a lifecycle of about 40 years. Ever since then the time left has continued to be about 40 years due to extra discoveries and production changes. The distinction between absolute and relative scarcity and abundance can best be seen through the work of Malthus and a contemporary David Ricardo. Malthus argued that land, for example, was a finite resource. Costs would be fixed until it had run out when costs would rise. Ricardo argued that there was enough land but only the quality (and hence profit on using it) varied. The good land went first. Poorer land was brought in later at a higher

cost (in much the same way as poorer grade metal ores are seen to be worth mining after profits rise).

5 **Pollution.** There has been a great deal of concern about pollution from energy usage (e.g. greenhouse effect/global warming).

Questions

1. If energy demands are related to the type of society what does this imply in energy supply terms for rapidly developing nations such as China?
2. Choose an energy resource e.g. gas. Using a range of resources, including contacting gas companies, describe the linkages of the energy system using the components of the cycle noted above. Compare this with other resources. What are the differences and similarities?
3. Using the examples of alternative sources noted above, use an outline map of the UK to suggest the best locations. Does any one area stand out as energy rich? Why?
4. Taking four or five examples of alternative energy. State where they can be found for viable usage. What is their conversion ratio? Could they make a significant contribution to energy needs?

2.5.8   References

Allen, J.E. (1992) *Energy Resources for a Changing World.* Cambridge University Press.

Brown, G.C. and Skipsey, E. (1986) *Energy Resources: Geology, Supply and Demand.* Open University Press.

Cassedy, E.S. and Grossman, P.Z. (1990) *Introduction to Energy: Resources, Technology, Society.* Cambridge University Press.

Chapman, J.D. (1989) *Geography and Energy.* Longman.

Hill, R., O'Keefe, P. and Snape, C. (1995) *The Future of Energy Use.* Earthscan.

Mather, A.S. and Chapman, K. (1995) *Environmental Resources.* Longman.

Wheeler, J.O and Muller, P.O. (1986) *Economic Geography,* 2nd. edition. Wiley.

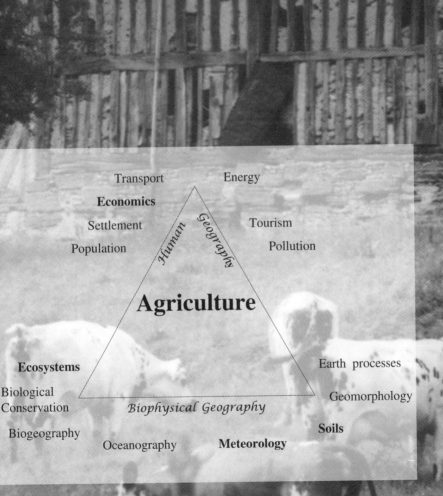

Transport   Energy

Economics

Settlement   Tourism

Population   Pollution

Human Geography

Agriculture

Ecosystems   Earth processes

Biological   Geomorphology
Conservation

Biophysical Geography

Biogeography   **Soils**

        Oceanography   **Meteorology**

Subject overview: agriculture

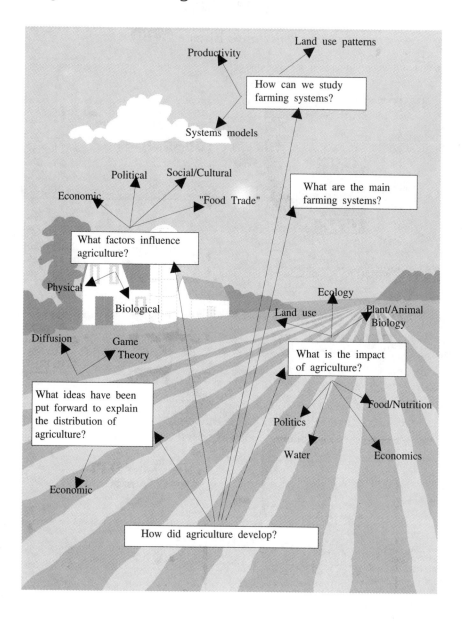

Productivity

Land use patterns

How can we study farming systems?

Systems models

Political   Social/Cultural

Economic

"Food Trade"

What are the main farming systems?

What factors influence agriculture?

Physical

Biological

Ecology

Land use

Plant/Animal Biology

Diffusion   Game Theory

What is the impact of agriculture?

What ideas have been put forward to explain the distribution of agriculture?

Food/Nutrition

Politics

Water

Economics

Economic

How did agriculture develop?

1. Farming started independently in seven distinct areas. Early farmers started from hunting and gathering to herding and breeding. Even today though, we rely on only a few species.
2. Farming ideas could spread by colonisation, diffusion or innovation.
3. Models explaining the distribution of agriculture can be divided into three – economic, diffusion and game theory. Economic models are the most popular and centre around the idea of changing profits with changing costs and distance from the market.
4. Numerous factors influence the actual distribution of farming types. These can be grouped under biological, physical, economic, political, social/cultural and 'food trade'
5. By using systems models we can compare and contrast farming types. Major groups are primitive/gathering, low input/Third world, collectivist and industrial.
6. The impact of agriculture depends on location and scale. Key areas include pollution, soil degradation, politics and the food trade, the food/aid issue, sustainability and global climate changes.

2.6.1   Introduction

The search for food is a basic need of all living things. Agriculture was one of the first ways in which people made a significant change to their surroundings. Even today, when we think that agriculture is no longer the force it was, we must recognise the importance of this activity on a global scale: nearly 50% of the world's population is involved in some form of agriculture or allied business (such as processing or selling). And yet, despite this, the whole idea is founded on a very thin base. Think of all the plant and animal species (low estimates reckon 5 million). How many are central to agriculture? About nine: wheat, rice, maize, millet, chickens, sheep, pigs, cattle, goats. One can add a few more to take account of more popular minorities such as potatoes and beans but even so, the list is minute compared to the number of wildlife species. What's more, these key species come from only two or three families. Perhaps the basic question of this chapter should be why we've stuck to such a small range rather than exploit a bigger wild supply.

2.6.2   How did agriculture develop?

Like many human economic activities, agriculture developed over several thousands of years. Using records from archaeological excavations and recent developments such as DNA analysis it is possible to gain some idea of the changes that have been made. There is evidence to suggest that certain pre-conditions would be necessary e.g. suitable plant stock, good existing food supplies (so experimentation wouldn't cause serious problems), water supply and a good reason to start in the first place. Given the stock the next stage would be to spread the idea. Possible mechanisms include colonisation (outside forces bringing change), diffusion (spreading ideas by word of mouth) and innovation.

Where did it happen? Figure 2.6.1 shows these areas and gives the approximate date when agriculture started. Combining a few of the regions we get five areas with their dominant food sources. The Fertile Crescent was noted for sheep, pigs, goats, cattle, barley and wheat. Europe and Africa had cattle, pigs, rice, millet and sorghum with East Asia utilizing rice, millet, pigs, chickens and buffalo. Mid and South America produced potato, maize, beans, squash, llama, alpaca and guinea pigs whilst North America had goosefoot and sunflower. Not all these are valuable today (goosefoot would be a prime candidate here). Note how few species are used and think about the implications of this. If we wanted to breed a new species we might want to go back to the original places. Often, these are areas of significant change/warfare e.g. Vietnam, Middle East. How do we safeguard such areas? Only recently has any attention been given to the conservation of domestic (as against wild) species and the idea is still in its infancy. Given that archaeologists have found over 2,400 species of maize alone then the biodiversity of agricultural species must be great.

Figure 2.6.1 Independent origins of agriculture

2.6.3 What factors influence agriculture?

There is a large range of factors that can influence farming and agricultural decisions. To make things easier factors can be grouped into six key areas: biological, physical, economic, political, social/cultural and the 'food trade'

Firstly, the **biological factors**. These include crop/stock type and their ecology. The aim would be to chose something that would grow well in an area and produce maximum profit i.e. using the optimum range ('best area' for a species). To this can be added the stocking ratio (number per unit area) and conversion ratio (amount of food compared with weight gained for animals and yield per hectare for plants) both of which are measures of the efficiency with which stock can use an area. For example pigs have a stocking rate of 2 per acre and a conversion ratio of 4:1 making them cost-effective in many places.

Another biological area becoming increasingly important is selective breeding and biotechnology. To increase production, many people are working on ways of breeding better stock and producing better plants. Some of this involves the use of hybrids (offspring not found in nature because the parents are dissimilar) whilst the very latest ideas are using biotechnology to produce completely new forms (or strains as they're called). All this is not without cost. The financial investment might well be beyond the poorer nations. There is also the risk of failure in the real world. Early **selective breeding** and **biotechnology** experiments in the **Green Revolution** produced very high yielding varieties that couldn't compete in real farming situations.

Secondly there are **physical factors**: climate, soils, relief and altitude. Every crop will have its optimal zone and the response to these zones on a global basis is a major factor in determining what farming types are seen where. As altitude increases so temperature decreases (as it does with latitude). This can best be seen in Figure 2.6.2 which shows how crops vary through the Andes. **Relief** can also affect the water supply and runoff in an area. This might mean areas of too little rainfall or that some areas get too much: waterlogging can be a serious problem (or it can be used to advantage: Medieval farmers used water-meadows' regular flooding to supplement the fertility). **Soil** (or **edaphic** factors) can also give rise to considerable variation in farming distribution. For example, whilst potatoes will tolerate a pH of 4.9, lucerne (for cattle feed) will need a pH of 6.2. Fertility is another important factor although this can be added to with artificial fertilizers or even farmyard manure. Despite these aids, areas of naturally high fertility (deltas, volcanic areas and fenlands) are highly prized and often grow the greatest yields in the region.

David Grigg argues that the **economics** of farming sets it aside from other enterprises because land is the prime requirement (rather than a site needed for a factory), external factors are crucial (such as nature), most producers are small-scale, most enterprises grow a range of products and

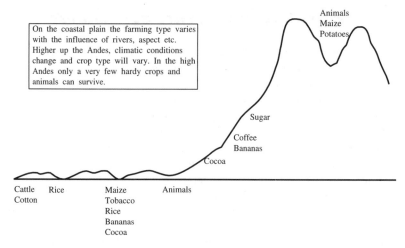

On the coastal plain the farming type varies with the influence of rivers, aspect etc. Higher up the Andes, climatic conditions change and crop type will vary. In the high Andes only a very few hardy crops and animals can survive.

Animals
Maize
Potatoes

Sugar

Coffee
Bananas

Cocoa

Cattle   Rice   Maize   Animals
Cotton   Tobacco
                      Rice
                      Bananas
                      Cocoa

Figure 2.6.2   *The influence of altitude on farming type*

that the social composition of many farms (i.e. as family units) is unlike that of industrial sites. When the farmer decides upon a particular activity there are ecological constraints but also economic considerations. There will be a certain amount of capital to invest and labour to employ. In addition, fertilizers, pesticides and machinery may be available. This can give rise to two 'extremes' on an economic continuum: **intensive farming** where the inputs (costs etc.) are high but so are the profits and **extensive farming** where the inputs are low and so are the profits (both in terms of unit area).

The intensive/extensive division might be useful but it can also be seen as too simple. For example, in Western Europe this picture can be altered by farms specialising in one crop/stock whilst others have a range of

Tip

Neither intensive nor extensive is the 'right' answer; it depends upon the quality of the land and other inputs in a given situation. We assume that the aim of any farm is to maximise its output (the difficult point is to define exactly what to maximise e.g. food, profit, work).

produce. Some farms may produce items that fall under the terms of the Common Agricultural Policy ($\Rightarrow$ European Union). Farmers may produce only what they like regardless of cost (e.g. cattle or pig farming). Despite such arguments most of the world's farmers are subsistence units growing just enough for family needs.

These ideas are only one part of the picture. Other economic aspects would include:

1 **Tenure** and **land ownership**. Farms may be owned as freehold (the farmer owns it outright), leasehold (the farmer rents it) or as common property (e.g. the former Soviet Union collectives and China). This will alter the flow of profits and therefore the development of the farm. In many parts of the developing world, land ownership is a major social issue. Where 10% of the people own 90% of the land as in some nations then social unrest and rural depopulation may follow (which causes problems in urban and rural areas).

2 **Farm size** can be seen an another variable. Studies examining the sizes of farms have discovered some remarkable patterns. Small farms (less than 5 hectares) dominate in Africa whilst very large farms (>1000 hectares) are common in Australia. This will affect the scale of the operation (and its ability to make economies of scale ($\Rightarrow$ economics). However, it's not just size that's important but also the distribution of fields: is the farm fragmented? One can imagine that the economics of running a fragmented farm is poorer than a consolidated farm which might explain the French government's moves to see farms consolidated in the 1950s.

3 **Marketing** is becoming more important especially in Western Europe where farmers work in co-operatives to get the economies of scale needed to compete. In Normandy, France, this includes cheese and wine producers whereas in the USA it might involve grain producers.

4 **Transport** increases the range of goods we can buy but can put pressure on weaker producers (such as in the Third World) to reduce selling prices. Ironically, it also reinforces our interdependence on each other because we can no longer rely upon just one area's produce.

5 **Labour supply**. In Western Europe, the number of people involved in farming has been declining for some time with the current figure at about 2%. This might make us see agriculture as less important (especially when it no longer dominates the economy) but we should not forget that there are vast areas where its significance to employment and the rural economy is huge. The Indian government has been so concerned about rural depopulation that it has tried to find ways of keeping people in the villages: one promising scheme involves concentrating any money on farming: keep the farmers going and the rest of the economy will follow.

Politics is an increasingly important factor behind agriculture. Hardly a day goes by without some media story of the European Union's Common Agricultural Policy (probably the most well-known and the most complex of the EU policy areas). The whole idea of subsidy and comparability has distorted a farming system (particularly in the UK) already distorted by local subsidies.

Social and **cultural** factors play a vital part in many areas of the world. Immigrants may bring with them techniques that are adapted to the new circumstances e.g. Scandinavian dairy ideas have been significant in the parts of the USA where they settled. Under the general heading of social and cultural factors we must mention **population**. It has been assumed in the past that as agricultural production rose so more people could be supported. In other words, the food levels rose before the population. There is an alternative view that has had profound effects on our examination of agriculture in the developed world. Esther Boserup, writing in the mid-1960s argued that as population density increased so would the intensity of the agriculture. In sparsely populated areas, shifting agriculture would be practised: few inputs and little labour required. With higher population density more food would be needed and so the land would have to be used more intensively Boserup saw population pressure as a reason for increasing agriculture. Her model was set up for African subsistence farming but it could be applied elsewhere.

The final key factor is the '**food trade**'. In the North, at least, the farmer is not a single individual but one part of a massive web of producers and consumers including processors, wholesalers and shops. The **linkages** (i.e. series of business relationships that one company would have with others) involved mean that the process has become highly complex. This can be demonstrated easily by reference to any local supermarket where shelves will be stocked with produce from around the world. Thus the food trade is made up of highly complex groups of links. It is also composed of individuals who each have their own political agenda. The list is large but includes farmers, farm workers, European Union administration, government departments (Ministry of Agriculture, Fisheries and Food etc.), pesticide/fertilizer suppliers (known as the agro-chemical trade), farming unions for workers, farmer unions (the UK's National Farmers Union is recognised as one of the most powerful lobby groups), commodity traders and the Stock Market (who trade in produce), food processors (like Nestlé and Unilever), wholesalers, shops/supermarkets, caterers and consumers producing a complex relationship for companies.

Studying the power of groups is not one of the first ideas of a geographer but this is changing fast especially as agriculture becomes just one part of the food trade. Studies in agriculture and industry are recognising the power of groups (such as multinational groups like Unilever and Nestlé). Their actions can distort markets and growing patterns. The traditional linkages in agriculture would have meant that a variety of organisations would be involved. Today, we see that many of the larger companies will control a series of enterprises throughout the food production chain.

2.6.4   What ideas have been put forward to explain the distribution of agriculture?

The factors mentioned above can be used to explain the distribution of farming types. The next stage is to use such factors to produce models

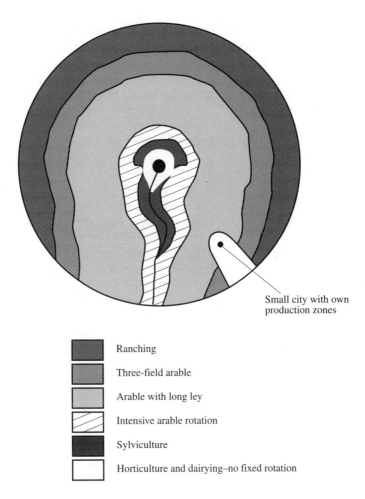

Small city with own
production zones

Ranching

Three-field arable

Arable with long ley

Intensive arable rotation

Sylviculture

Horticulture and dairying–no fixed rotation

Figure 2.6.3 Von Thünen's model

which might help us understand which factors are the most crucial. Most ideas use location models: they attempt to explain why a certain farming system is found where it is. There are three main groups of models: **economic, diffusion** and **game theory**. Economic theories were the first to be put forward and still hold considerable power. The first of these was devised by a German economist and landowner, J. H. Von Thunen whose book, *The Isolated State*, was published in 1826. The basic idea behind this was that the location of any given type of farm depended upon its distance from the market. At any one place, the profit (P) made from the sale of produce would be the sale price (S) less the cost of production (C) which gives the simple term:

$$P = S - C$$

The only cost that would vary would be the cost of transporting the goods to market (T). As the distance increased so the profit decreased. Eventually there is a point where there is no profit to be made (see Figure 2.6.3). For each crop there is a different slope to the line. This means that the transport and production costs differ. Which type of produce is found where? At any one point the topmost line gives the greatest profit. Thus closest to the market 'a' is dominant; where 'b' crosses over it becomes the most profitable and so on. This idea has been subjected to a series of comments and modifications over the years but the basic idea still holds: produce will only be grown where a profit can be made: in an ideal situation the produce giving the highest profit (return) will be farmed.

One other model has been devised using similar ideas but actually comes out with a completely different answer! Using the USA as an example, but keeping to the idea of profits, Sinclair produced a model where intensity would increase away from the market (see Figure 2.6.4). The urban farming zone would be less well used. It would be no good investing in an area that might soon change e.g. for urban expansion. Likewise, the vacant, grazing and field crops zones would be subject to change. Only when far enough from the market would you be able to guarantee production and therefore justify investment.

Economic models are powerful because agriculture is an economic activity. They have less power when describing situations where profit is not the motive such as in subsistence farming. This means that there are many developing nations where these ideas do not provide good explanations.

Innovation diffusion models offer an alternative perspective. The argument here is that agricultural distribution is caused by the spread of ideas from one farmer to a friend (**social diffusion**). Alternatively, one farmer could start and the neighbouring farm could join in, thus spreading the idea through the district (**spatial diffusion**). Finally, **game theory** uses mathematics to produce a theoretical distribution of ideas which can then be subsequently be checked by fieldwork.

> **Tip**
>
> Models aim to find common factors which can explain variations in farming patterns. They don't try to account for every variation.

2.6.5 What are the main farming systems?

Of the many methods used to classify and describe farms one of the best is the systems model (⇒ models). This has the advantage of allowing us to compare inputs and outputs as well as see how much energy is used to farm the system. Also, it can be used on a wide range of farming types. The

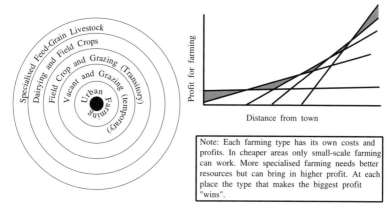

Figure 2.6.4 Sinclair's model

Note: Each farming type has its own costs and profits. In cheaper areas only small-scale farming can work. More specialised farming needs better resources but can bring in higher profit. At each place the type that makes the biggest profit "wins".

major problems come when trying to get the data needed for a detailed analysis especially when looking at systems over time. This means that specific examples have to be studied and the generalisations taken from these. The aim here is to compare four key farming systems: New Guinea 'gardening' system, an Indian farm, a Russian collective farm and a modern UK farm.

The first case is the New Guinea 'gardening system' – a small-scale farming operation based on minimal inputs and a shifting field system. This can be seen as a 'simple' system in terms of European agriculture but with recent concerns about the environment perhaps a better phrase would be low impact. The farms in this example are situated in the highlands of New Guinea where a mixture of slope (in excess of 20 degrees usually), high rainfall, montane rain forest and poor soils mean that **shifting agriculture** (also called **swiddening**) is practised. This means that when the soil starts to lose its fertility, a new area in chosen. Principal crops are cabbages, onions and sweet potatoes whilst the pig is the dominant stock. The chief input to this system is human labour. Goods produced are used by the villagers with the exception of a little exchange of surpluses (which also has a social value). A systems diagram (see Figure 2.6.5) shows the key features. One of the problems in New Guinea is the maintenance of fertility. Shifting areas is one way: ensuring that nutrients are constantly brought into the farm area is another.

The next case is an Indian wet-paddy rice farm. Here, dry land is converted into an aquatic system through the use of irrigation. Fertility can be maintained because the nutrient supply is constantly refreshed. In addition to nutrients, constant waterlogging changes soil structure to make it impervious i.e. water retaining. Waterlogging encourages the growth of algae which contain nitrates for rice plants. This allows the land to grow a greater yield of crops. So, with rice increasing the yield, the farmer can use some land for other purposes. Sugar cane grows well in many areas whilst

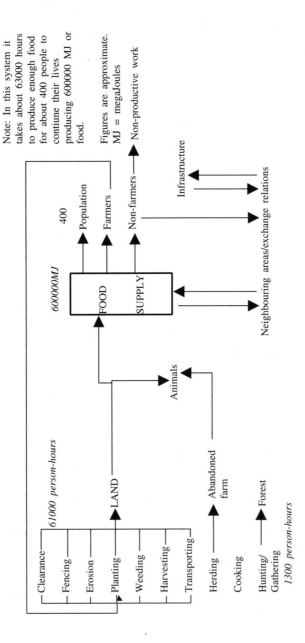

Figure 2.6.5 Energy flow through a subsistence farm

millet can supplement rice as a main food source and any ground remaining can be used for bullock grazing (the other source of power). One can see what this does to the energy flow system in Figure 2.6.6.

The current political situation means that there is less emphasis on the old Socialist countries like Russia. Yet there too were farming systems which could be used to demonstrate changes in agricultural production. The main change was that of **collectivisation**: the ownership and organisation by the State which operated at farm level through the collective. The typical farm in the Moscow region would be almost 50:50 grassland and arable. This would allow a rotation system to be developed where a three-year wheat–barley–fallow pattern would be developed. This is another way of overcoming the nutrient problem (recall how it occurred in New Guinea and India). In addition, other crops such as cabbages and potatoes would be grown. To increase energy input, tractors and combine harvesters were used. The impact this would have on the system can be seen in Figure 2.6.7. Another impact, not so easily seen, is the influence of central planning. The idea of collectivisation was to improve the efficiency of Russian farming (which was a small-holding system with few industrial inputs). In so doing, the public demands could override the private interest and so a public:private split would occur. Often, yields on the private plots (which would still be allowed to provide farmers' food) are quoted as being far higher than on public land. Even now, this is subject to debate so it is best to say that this system is another example of the spectrum which takes us from hunter to industrial farmer.

This final example is from Southern England. The system that existed in Victorian times has been transformed. Thus the 'typical' English farmland has seen changes as great as in any other system. Major changes include rural depopulation (farm workers moved to towns), mechanisation from the 1940s onwards, increase in chemical inputs (fertilizer and pesticides) and amalgamation of farms. Here there is an example of the extent to which ecological efficiency in farming could be taken. How could this be achieved? Recall the other systems. Each tried to improve on the other by the increase of inputs. Shifting agriculture could be 'outcompeted' by wet-paddy rice. Collective farms with their machinery input would bring greater yields. To achieve the highest level would need a considerable increase in both fertilizers (artificial rather than natural) and machinery. This has led to the situation where current agriculture actually consumes more than it produces (in energy terms)! If we look at the output:input ratio for, say, dairy farms the figure of 0.4 means that for every 10 units of input we get only 4 units of produce.

2.6.6   What is the impact of agriculture?

Increasingly, the impact of farming systems on the environment is being

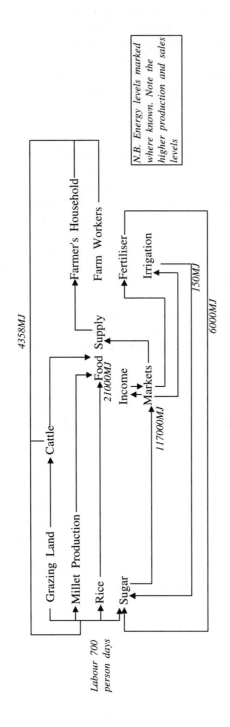

N.B. Energy levels marked where known. Note the higher production and sales levels

Figure 2.6.6 Energy flow through an Indian farm

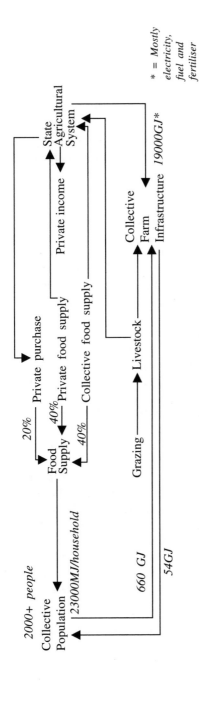

N.B. *Energy figures marked where known*

Figure 2.6.7 Energy flow through Russian collective farm

questioned in Western. Europe and North America even though in many (if not most) nations the key issue is the production of enough food for the population. Issues in farming have changed as any examination of the farming journals over the past 20 years would demonstrate. Today, the key issues are:

1 **Pollution** – farm pollution especially from silage (grass) is the biggest single cause of water pollution. Although this is serious in areas with a high stock rate a similar problem can be seen in East Anglia where fertilizer run-off can reach levels too high to drink tap-water safely.

2 **Soil degradation** – potentially a major global problem. If too much land is affected especially in areas difficult to cultivate anyway then it could affect large numbers of people (especially South of the Sahara Desert).

3 **Agricultural subsidies** – a major issue with the European Union. The imbalance this creates has been open to abuse. It has made food a political issue as well as distorting the economic value of produce.

4 **Food multinationals** – farming has gone well beyond the farm boundary. With an increasing number of multinational companies buying a range of food companies (from farms to packaging) their influence has grown. The impact of this has been a source of controversy especially in some cases e.g. baby milk powder and its marketing to developing nations.

5 **Food and aid** – should one feed people because they are hungry or because their political system might support yours? Does feeding help or should we be helping to establish agriculture? Does food aid help or wreck the local farming economy? Just three questions raised in this highly charged area.

6 Is modern farming sustainable? – **sustainability** is one of the in-words of the 1990s. Unfortunately there is no complete agreement on what it means or how it can be achieved. The basic idea seems to be achieving enough food but using less harmful ways of doing it.

7 **Surplus land** – with many areas short of food in Western Europe we have cases of land being taken out of production because it can be worth more through subsidy that way. Alternatively, other uses can be found for the land (such as golf courses). How do we plan land?

8 **Global climate change and food security** – if climate changes how will it affect food production? Will some nations currently in surplus be in deficit in a few decades' time? How will nations react to food shortages? It's interesting to note that questions like these are being studied not just by agriculturalists but by government security advisors!

These are just a very few of the issues from local cases to global issues. It shows that farming is still a key part of our lives. We need to study it in detail to appreciate its impact.

1. What would happen if all wild ancestors of domestic species were allowed to die out?
2. What flaws can you see in Von Thünen's original argument? How is this complicated by modern farming practices? Can you think of any situations where a farmer may keep with a less profitable farming type?
3. Look at the types of farming carried out around your town? Does either Von Thünen's or Sinclair's models fit what you can see? What causes any variation?
4. Compare and contrast the four systems mentioned in the text.

2.6.7 References

Baylis-Smith, T.P. (1982) *The Ecology of Agricultural Systems*. Cambridge University Press.

Briggs, D. and Courtney, F. (1989) *Agriculture and Environment*. Longman.

Grigg, D. (1995) *An Introduction to Agricultural Geography*, Second edition. Routledge.

Smith, B.D. (1995) *The Emergence of Agriculture*. WH Freeman.

Tansey, G. and Worsley, T. (1995) *The Food System: A Guide*. Earthscan.

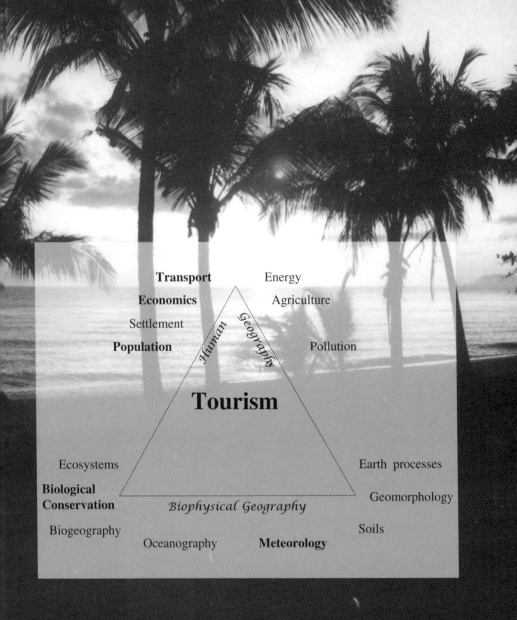

Transport   Energy

Economics   Agriculture

Settlement

Human Geography

Population   Pollution

Tourism

Ecosystems   Earth processes

Biological Conservation   Geomorphology

Biophysical Geography

Biogeography   Soils

Oceanography   **Meteorology**

Subject overview: tourism

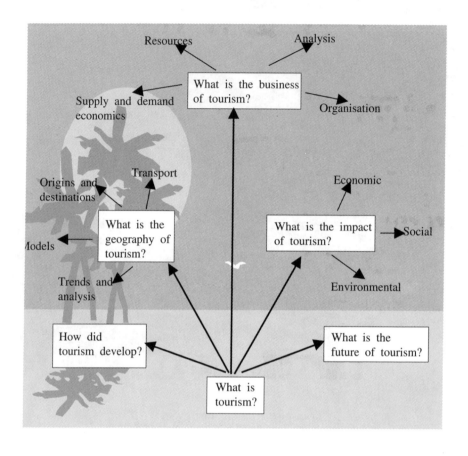

1. Many concepts like tourism and recreation are difficult to define because there are so many variations and it's based on personal choice and perception.
2. Tourism has a history stretching back hundreds of years but has only become universal in the twentieth century.
3. Tourism has spread by the ripple effect with innovators visiting new areas and mass tourism following. This has been made possible by increases in leisure time and money and availability of transport.
4. The geography of tourism can be understood by studying four concepts: trends, models, resources and impact.
5. The key trends in tourism are its growth as a leading economic sector and its spread to all members of society.
6. There are four groups of models commonly used to explain tourist geography: travel-flow, origin-destination, structural and evolutionary models.
7. The study of tourism is divided strongly into geography and business. Business studies the economics, resources, organisation and strategies of tourism companies. A key idea is the elasticity of tourism; it can, and does, change rapidly.
8. The impact of tourism is both positive and negative. It covers aspects of social, cultural, economic and environmental concern.
9. As tourism grows it's in danger of destroying the very features it sells. Ways forward include 'sustainable tourism' and 'ecotourism'. The future is not clear cut.

2.7.1 Introduction

Although tourism is one of the world's fastest-growing employment sectors and in some places it provides the only real income for the local people, tourism geography is one of the newer subject areas. Given the increase in leisure time and the rise in incomes more people are deciding to pursue leisure activities. The impact that this is having on social patterns, economic organisation and the environment makes it crucial that we study this field.

2.7.2 What is tourism?

It might appear obvious at first but in trying to actually answer the question it becomes less easy – there are many difficulties. The aim of this section is to introduce some of the key ideas. To assist, the following definitions will be used in this chapter:

1 **Leisure** – time when you are not involved in employment. Of course this raises the issues of unemployment (is all that leisure time?) and those who are employed but not paid (e.g. students, voluntary workers etc.). It is possible to improve on this definition by saying that it's the time left after all the necessary work is done (e.g. working, sleeping, commuting etc.).

2 **Recreation** – any activity done for pleasure in leisure time without necessarily expecting any money. This means that you can follow a work interest as a hobby (e.g. computing) and that you could try to make money from the activity (e.g. photography) but the emphasis is still on personal choice. In practice, the choice of recreation depends on your personal interests and your ability to obtain it; recreational activities are dependent upon your income. This means that some hobbies e.g. golf, have a limited range of socio-economic classes involved.

3 **Tourism** – recreation requiring at least one night's stay away from home following a recreational interest. Again, there's the problem of the business person who, in addition to carrying out some work, also spends a few day's vacation in a place. As with recreation, the choice of tourist area depends on your ability to pay for it.

Apart from the money and personal interests mentioned in these definitions there is also the amount of time available. If there's only a few hours then people stay close to home for recreation. With a few days it's possible to travel further. In theory it's possible to divide any area into local zone, day-trip zone and tourist zone (together, called **trip-zones**). If this were done with another person's home you'd find that the circles would all cross. This means that any facility used in your town would be a local visit but for someone else it would be a tourist facility because they would need to stay away from home to see it! Looked at this way any recreational site can be seen as a tourist facility because it could be used by tourists. Leiper made a simple tourist model to explain this situation (Figure 2.7.1). Tourists would come from a **generating region** and **travel** to a **destination** – the three elements that Leiper considered important.

It's possible to add another dimension to this by assuming the tourist

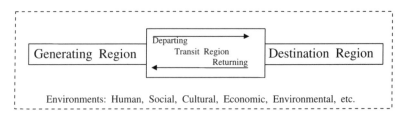

Figure 2.7.1 Leiper's tourism model

system is surrounded by **personal choice** which would be your decision based upon a range of factors including age and gender; education and interests; socio-economic group, occupation and status; family and stage in the life cycle; time available; distances involved; purpose of visit; transport available; and facilities needed. Although the choice is, in theory, infinite once you come down to it there are far fewer possibilities.

2.7.3 How did tourism develop?

Tourism is a not recent activity. It has been going on for thousands of years: it's just that it has been very restricted until recently. The problem is that tourism needs time and most people didn't have any free time. One of the earliest examples of tourism would be the pilgrimages of Medieval England. The idea was to visit Holy sites with the aim of becoming a more religious person. At the same time, abbeys and monasteries became early 'hotels' giving food and lodging to visitors. As today, the richer visitors got better treatment. Also like today, you could get souvenirs – pilgrim badges – in lead, silver or gold. Some even bought several to trade with other pilgrims. We even have records of an exchange rate with a Jerusalem badge being worth several Canterbury badges!

Leaving aside the religious visits the idea of tourism as we know it started in the seventeenth century in England. Two forms could be seen. Royalty went to a select number of sites to 'take the water' – literally to drink certain mineral waters in the hope of restoring health. Tonbridge Wells and Bath are just two examples. The other form was the 'Grand Tour' where wealthy people toured around the sites of ancient Greece and Rome to improve their education. By the eighteenth century both forms had become popular and other people tried to join in. For example, once it became fashionable to go to Bath, other wealthy people came which led to the Royalty moving to newer sites such as Brighton. In the early nineteenth century Mr Thomas Cook started arranging tours for people wishing to see Europe. By the late nineteenth century the railways had started to have a considerable effect on people's recreation. By adding an ability to travel to a small (but slowly increasing) amount of time and money, mass tourism started to grow. Initially, this was just a day trip to the seaside but soon other areas were opened up. At the same time, the richer people moved away from the coast to French resorts such as Deauville.

In the twentieth century the tourist demand built up considerably. Road transport meant more places could be visited. Paid holidays of a week or two meant that time and money was available. The 1920s saw the peak of motor bus travel and rail holidays. By 1950 the car started to make a great impact. By 1960 the aircraft industry could produce good, reliable transport and the UK 'package tourist' industry was born. In this type of holiday, all the travel and accommodation and even food was put together as one item. Virtually all of this activity was for foreign holidays with Spain as a major destination. Subsequent changes have really been extensions of this initial idea. For example, car-based holidays are now very well catered for with facilities not just in the UK but also in France and elsewhere in continental Europe. The 1980s saw the rise of international tourism outside Europe with the Far East and the USA experiencing growth in tourism. Of course, the reverse is also true with the UK seeing increased numbers of US and Japanese visitors.

Although this has only been a very brief guide to tourism, there are some crucial points that must be emphasised:

- Tourism can be seen as developing by waves or ripples. There have always been the 'pioneer' travellers trying new places and experiences (Bath, Brighton, Deauville, Switzerland, Greek islands, Australia taking the UK case outlined above).
- The growth of tourism has been so great that it has had a considerable and usually unplanned impact. In some places the costs have outweighed the benefits and there is evidence to suggest that a redevelopment could take place e.g. in Spain, mass-holiday destinations are being rebuilt to cater for wealthier tourists in lower density accommodation.
- Tourism presents a challenge to geographers because of the complexity of its nature and the difficulty of obtaining data. In product-cycle terms, tourism is still in the innovation phase. There's still considerable development although places such as the UK with a longer tourist history are showing development of new products to take over as old ideas fade. (e.g. the 're-invention' of the holiday camp to provide the broader experience demanded by current holiday-makers). To some extent the fall of the seaside holiday and the rise of the package trip are examples of this.

2.7.4 What is the impact of tourism?

There is no denying that tourism has an impact (both positive and negative) and that it requires careful analysis. Although its been suggested that impact is universal it's effects are not always felt. This is because the area under study might be able to withstand the impact without changing. The ability of an area to withstand impact without damage is called the **carrying capacity**. Most writers argue for four different carrying capacities:

1 **Physical** – the amount of damage the area can withstand without changing. Of course this also depends on the number of people and the amount of time involved – a weekend gathering can cause more damage than a trickle of visitors throughout the year.
2 **Psychological** – this carrying capacity is exceeded when a visitor feels that the experience is made poorer (e.g. by too many people). This is highly personal and depends upon what the person likes.
3 **Biological** – the ability to withstand stress without changing. In this context, the capacity is exceeded if the plant and animal composition changes.
4 **Social** – a carrying capacity which defines the limit of tourism acceptable to both visitors and residents.

These carrying capacities are largely personal. This both reinforces the idea of tourism as a highly individualised activity and creates difficulties in trying to research it. Having seen the range of ideas in assessment the next stage is to examine some impacts in more detail. Three are crucial: environmental, economic and social.

The **environmental impact** of tourism is one of the main talking points today. Since one of the main reasons for visiting a new place is to appreciate the scenery then its quality is obviously going to be a major factor. Often there is the assumption that impact is going to be negative. Whilst this is the more common direction for tourism there is also a positive aspect. For example, urban settings might provide more nest sites and have good food supplies. Some newer tourist business developments (usually referred to as eco-tourism) are actually encouraging a positive approach. Given that enjoyment of wildlife, fishing etc. is a major hobby activity it follows that concern on both sides is going to be great. One framework that allows a closer examination of the problem is seen in Figure 2.7.2. Although there is the assumption of negative impact, it does highlight some of the damage that can be caused and also demonstrates the difficulties in providing truly environment-friendly ecotourism.

If the impact of tourism on the environment is controversial then the **economic impact** is equally so. A common approach is to use some form of **cost-benefit analysis** as seen in Figure 2.7.3 which highlights not only the costs and benefits to those directly involved but shows the breadth and depth of community involvement. The issue is not just about the spread of economic impact but its distribution – who gets the most? There is evidence to suggest that the costs and benefits are not equally distributed. Seasonal employment and low wages characterise employees whilst developers can make a great deal of money. This might even go abroad if the company is a multinational or involved in vertical integration. If all this should be seen as negative then there is the positive impact on the economy

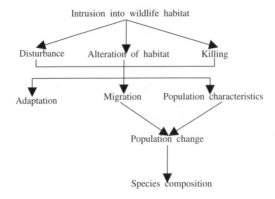

Figure 2.7.2 *Environmental impact of tourism*

as shown by the **multiplier effect**. Briefly, this states that every pound spent generates more income to the local economy. Thus the hotel will gain money from the tourist which it spends buying local goods. The goods supplier will pay staff who will also buy local goods etc.

Finally, there is the **social impact** of tourism. Table 2.7.1 highlights some of the key impacts. It's worth considering this list in a range of contexts. It assumes that the impact will be greatest where the differences are greatest. Thus UK tourists in the UK are going to cause less impact than the same number of Britons in, say, Brazil.

2.7.5 What is the business of tourism?

Unlike other aspects of economic geography, tourism has developed two very strong and distinctive sides to its study: geography and business. The geographical part deals with the distribution of tourist facilities (see section 2.7.6). Any organisation or facility has to produce some form of profit (even if it's just community satisfaction with a local park that people are prepared to pay local taxes to keep up). Figure 2.7.4 shows some of the key

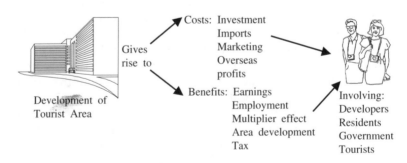

Figure 2.7.3 *Costs and benefits of tourism*

Table 2.7.1 Social and cultural impacts of tourism

- Impact on the population structure
- Rural to urban migration
- Changing job structures and markets
- Impact on qualifications
- Demand for female labour
- Seasonality of employment
- Changing moral and cultural values
- Changes to traditional way of life
- Changing structure of arts and crafts
- Changing consumption patterns
- Benefits to the tourist – relaxation etc.
- Impact on knowledge base of host and tourist

aspects of the **economic environment** of the business. On the demand side a key aspect is personal choice. **Consumer theory** helps one to understand why people choose what they do (Figure 2.7.5) and shows you how complex choice is. Furthermore, because personal choice can change rapidly it's possible to say that demand for tourism is 'elastic' i.e. any change in price (or other variable) can cause considerable change in demand. This means that the tourism business is very susceptible to changes in public opinion. On the supply side, the costs are often fixed – any change in demand can cause heavy losses. Alternatively if the demand is fixed and the prices rise losses can also occur. These are two reasons why some writers feel that an economy based on tourism is not as stable as one based on manufacturing. It also explains why market forecasting and other research tools have such a crucial role in tourism economics!

The resources of an area play a vital part in attracting tourism. They also determine what sorts of tourism are viable. This has led some writers to classify recreation resource areas into three: user-oriented (usually built facilities like golf courses), intermediate (natural areas with heavy user pressure) and resource-oriented (remote, natural areas for sightseeing

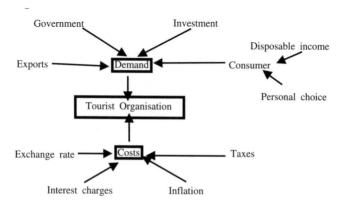

Figure 2.7.4 Economic environment of tourism

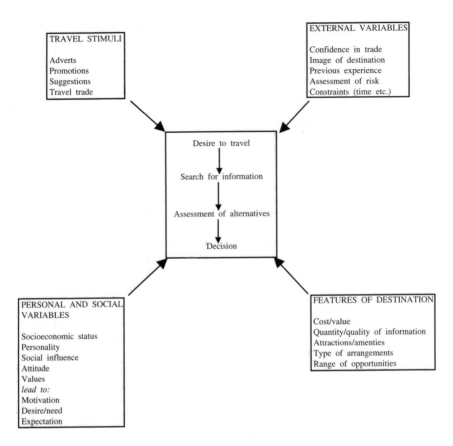

Figure 2.7.5 *Factors in tourist customer choice*

etc.). Taking these three as a series then the pressure of use decreases and distance from population increases from user- to resource-area. The former is artificial and activity-oriented; the latter is natural and environment-oriented. There's the additional point that user-oriented areas tend to be the core whilst resource-oriented areas are often the periphery.

Because of its diversity, the organisation of tourism produces a complex picture. Not only is there a range of different organisations there are also groups of organisations linked to provide a more profitable service. Two groupings are common: horizontal and vertical integration. Horizontal integration could be an airline buying other airlines and routes to maximise its business. Vertical integration occurs when a company buys a chain of businesses each contributing one part to the tourist facility. One common example is where an airline buys hotels, buses, catering firms, tour operators etc. so that in theory the tourist may never pay anything to a locally owned tourist facility.

In many ways the study of the geography of tourism is like any other economic activity. The aim in this section is to study trends, models, the nature of the resource and its impact.

Understanding the **trends** behind the growth of tourism is one of the most important functions for any tourism business. Tourism (and leisure and recreation) is based on personal preference. This distinction between it and other sectors is crucial. Although there is choice in car manufacturers the person is still going to buy a car. Thus steel manufacturers only need to know where to send the steel. Holiday firms have no idea whether someone is going to buy a holiday and if so, where they are going. Recent trends in foreign holidays have shown a dramatic swing from cheap package tours to more expensive holidays. One result has been an oversupply of cheap holidays and the considerable loss of revenue to places like Spain.

Most businesses want to maintain or increase their market share. In tourism this means keeping up with the ways in which people think. If a company pursues one tourism product and customers want another then **strategic drift** is said to have taken place i.e. the company's strategy and that of the holidaymakers is at variance. One way of assessing trends is to use two forms of business analysis: 'PEST' (political, economic, sociocultural and technological environments) and 'SWOT' (strengths, weaknesses, opportunities and threats). For example, an airline could use both forms of analysis (as outlined in Table 2.7.2) to assess its strengths and weaknesses (SWOT) and set this against the changes in the sector (PEST) and a study of the 'competitive environment' (the direct competition against the company).

Tip

Although this analysis has been using a company as an example there's no reason why it couldn't be a regional or even international comparison.

Whereas the PEST/SWOT analyses have been part of the business environment it is also possible to devise more conventional geographical models to help explain major trends. This is the realm of **models**. As the subject has developed so an increasing range of models has been made. Many are highly complex and would be difficult to work out in practice. However, the basic ideas behind them are useful. To assist, models can be divided into five categories depending upon the factors used:

1 **Mathematical models** – have been made to examine a very limited range of ideas. One, the gravity model, has been used in many instances. Here it's assumed that the importance of an area increases

Table 2.7.2 PEST and SWOT within a competitive environment

Environment	Strengths	Weaknesses	Opportunities	Threats
Competitive:				
Entrants				
Buyers				
Suppliers				
Substitutes				
Competitors				
PEST:				
Political				
Economic				
Sociocultural				
Technological				

with its size and nearness. Thus the tourist flow (T) between two places A and B is:

$$T = k(Pa \times Pb)/Dab$$

where k is a constant; Pa and Pb the populations of A and B, respectively and Dab the distance between them.

A more interesting measure is Defert's Tourist Function Index (TFI). This is very similar to ideas of location quotient where we are looking for concentrations. Here:

$$TFI = (N \times 100)/P$$

where N = number of tourist beds in an area and P its resident population.

2 **Travel-flow models** – examine the movement of tourists from origin to destination. Several examples of increasing complexity can be found. This example shown here (Figure 2.7.6) distinguishes both different types of travelling and introduces the key geographical ideas of origin and destination.

3 **Origin-destination models** – study the impact of tourists at both the

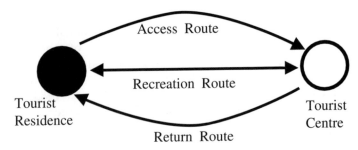

Figure 2.7.6 Travel-flow model

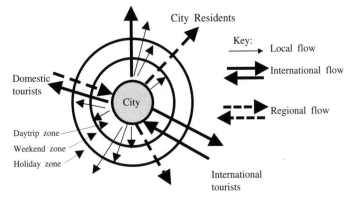

Figure 2.7.7 Pearce's model of tourist flows

beginning and end of their journeys. Various examples can be seen of which Pearce's (Figure 2.7.7) is the more complex.

4 **Structural models** – assume that tourism is part of the economy. Britton's enclave model is a typical example (Figure 2.7.8). Here, the business is controlled by powerful organisations at the origin. Tourists buy their holidays through a restricted number of outlets (e.g. travel agents) controlled by a few multinationals (e.g. British Airways). At the destination, the control, still with the origin-company, will mean visitors stay only in certain places (enclaves).

5 **Evolutionary models** – attempt to bring the element of time into the equation.

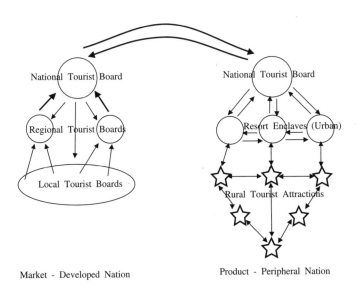

Figure 2.7.8 Structural model: Britton's enclave model

The third aspect of this study refers to the **nature of the resource**. Using the same ideas as seen in industrial geography the tourist area becomes the 'raw material' or resource base of tourism. Three different ways of classifying this resource base can be seen in Table 2.7.3. Rather than classify areas, some workers have tried to see how an area can change over time. Using ideas similar to the business cycle, the tourist area life cycle shows how a place might develop through time (see Figure 2.7.9). Another approach has been to assess the potential of an area for tourism. Duffield and Owen use four factors: suitability for land-based recreation, suitability for water-based recreation, scenic quality and ecological significance. By scoring each of the four according to pre-set criteria and weighting the result it's possible to get a score for each grid square on the map. The resulting pattern highlights areas of potential.

Table 2.7.3 Classifying tourist resource bases

Lavery	ORRRC	Clawson
Capital city – high standard, short-stay, international	**High density recreation area** – wide variety of use, extensive facilities. Intensive use	**User-oriented** – Based on resources available. Built environment. Intensive, heavy use of facilities
Select resort – high and lower standard accommodation. Attractive scenic area	**General outdoors area** – wide variety of use, substantial development	**Intermediate** – Best resources. Access important. High degree of pressure and wear. Often considerable scenic resources
Popular resort – wide range of accommodation, highly seasonal	**Natural environment area** – multiple use area	**Resource-based** – Often remote. High quality environment with low-intensity development
Minor resort – limited tourists and facilities, less accessible and popular	**Unique natural area** – scenic resource, sightseeing	
Cultural/historic centre – high proportions of international tourists	**Primitive area** – untouched wilderness	
Winter-sports resort	**Historic/Cultural site** – buildings, monuments, shrines, spas, pilgrimage sites archaeological area	
Spa		
Day-trip resort – close to and dominated by local population		

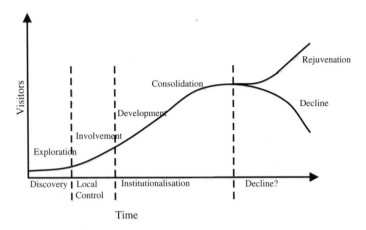

Figure 2.7.9 Tourist area life cycle

The final aspect in this section is the **impact** of the tourist industry on other sectors of the economy. Given the importance of transport in tourism it seems appropriate to find out how tourism can affect it.

From the earliest days of tourism, transport has been a key theme. In the nineteenth century this was the railway system. In the twentieth century road (initially public and later private) and air transport have come to the fore. One of the more interesting features about tourist travel is the way it has affected transport systems. Perhaps the best example of this is the airline business. Two examples illustrate the type of changes: **freedom of the air** and the **hub and spoke system**. When planes first started to fly their use was mainly military and no agreements were needed. The rise of civil aviation meant that some regulations were needed to keep the system working whilst still allowing nations some say over their airspace. The 'freedom of the air' agreements allow aircraft movements that would not be granted for road transport. These freedoms include rights to fly over nations, stop, pick up people/goods etc. These might seem simple but think about it: the landing of aircraft on someone else's runway could be considered an act of war.

The second example is the hub and spoke system. Economic theory suggests that location be made at the least cost site. With aircraft pricing that can be a difficult matter: some tourists want to stop at a large city whilst only a handful want the small town. By using a hub and spoke system an economic service can be produced. All big airlines have key airports. For example the US carrier Northwest has its hub in Minneapolis/St. Paul. It you want to fly on to Madison, Wisconsin then you'll need to change. If you came from the UK you'd overfly Madison but because the traffic doesn't justify it you'd have to fly back (the pricing structure is such that it's often cheaper to do this than find a direct flight).

2.7.7 What is the future of tourism?

Where does all this leave the future of tourism? Consider the main aspects of the sector as outlined so far in this chapter:

- considerable growth throughout the twentieth century
- development of a sophisticated business sector
- change in work/leisure patterns (for both those able to take holidays and those who supply them – but in different ways!)
- growth in the range of tourist opportunities offered/demanded
- increase in impact of tourist operations

Two ideas stand out: the sector is increasing in size and sophistication and so is its impact. There is every indication that both of these will continue such growth at least in the short term. This is a worrying trend because it suggests that the pressure on resources is going to destroy the very thing that people have come to see. For example, pressure in the Great Barrier Reef Marine National Park in Australia is such that some areas are dying. Too many swimmers/divers near to sensitive coral soon turn the multi-coloured areas into a ghostly grey.

This rather negative aspect of tourism is already being countered. Recent ideas of 'ecotourism' or 'sustainable tourism' are gaining ground. Although there is a range of ideas under this heading the basic concept is to produce a minimal impact upon the destination. Table 2.7.4 shows some of the ideals that should be followed. As the sector is developing so is the demand for more individual holidays. The decline in mass tourism to Spain has already been mentioned; what we're suggesting here is the change of holiday intentions from mass to individual (usually called segment or niche markets).

This last point is part of the more positive side of tourism. Here, the economic potential of the sector is recognised and so the aim is to maximise that whilst still retaining the culture of the area. Core areas might provide the current wealth of the regions (e.g. EU) but the next wealth-producing area might be the tourist-oriented periphery! Such an idea is not so far-fetched. There is current concern about the way in which devel-

Table 2.7.4 Ecotourism guidelines

- Minimise environmental problems in all tourist products
- Be aware of environmental issues at all times
- Be sensitive to conservation of fragile ecosystems
- Reduce energy demands
- Recycle wherever possible
- Control waste disposal
- Minimise noise levels
- Eliminate environmentally unfriendly products
- Respect the culture and history of tourist destinations
- Consider environmental concerns as the key aspect in tourism

---

oping countries are being exploited for their tourist value and their culture being degraded at the same time. If the tourist flow could be stabilised (and there is danger in any economic operation when it is not) then it would be possible to base developing nation economies upon this tertiary/quaternary sector base (which would open up a whole new set of ideas).

Questions

1. What are the advantages and disadvantages of vertical and horizontal integration in the tourism business?
2. Make your own copy of Table 2.7.2. Taking one of the following examples, complete the table. What conclusions can you draw from your analysis? (Examples could be two airlines, two neighbouring holiday resorts, two national tourist agencies, rail vs. car for holiday transport etc.)
3. Describe the social, economic and environmental impacts that tourism has had on your local area. Compare this with another destination overseas. Are there similarities? Why?

2.7.8   References

Boniface, B.G and Cooper, C. (1994) *The Geography of Travel and Tourism*, 2nd edition. Butterworth-Heinemann.

Cooper, C., Fletcher, J., Gilbert, D. and Wanhill, S. (1995) *Tourism – Principles and Practice*. Longman.

Foster, D.L. (1994) *An Introduction to Travel and Tourism*, 2nd edition. McGraw-Hill.

Pearce, D. (1989) *Tourist Development*, 2nd edition. Longman.

Tribe, J. (1995) *The Economics of Leisure and Tourism: Environments, Markets and Impact*. Butterworth-Heinemann.

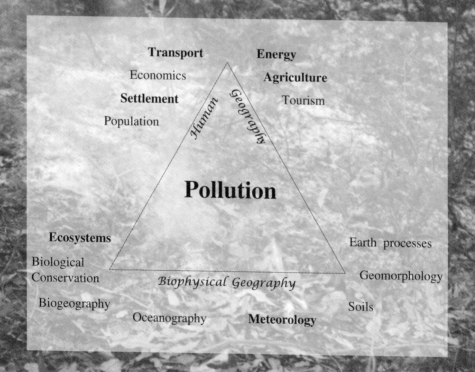

Transport

Energy

Economics

Agriculture

Settlement

Tourism

Population

Human Geography

Pollution

Ecosystems

Earth processes

Biological
Conservation

Geomorphology

Biogeography

Soils

Oceanography

Meteorology

Biophysical Geography

Subject overview: pollution

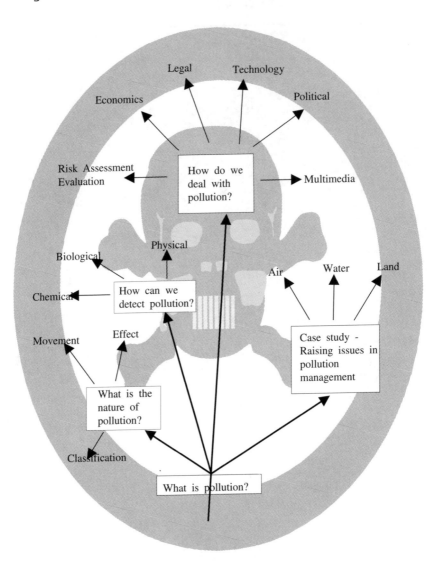

1. Pollution is a byproduct of most human activity.
2. Pollution is difficult to define but is usually taken as being too much of a substance (or energy) in the wrong place at the wrong time.
3. Most pollutants have natural counterparts.
4. Pollutants are produced at a source, travel by a series of pathways through the environment to reach a target.
5. The impact of a pollutant depends upon its persistence, lifetime, interactions, speciation, pollution loads, dispersion, ability to be concentrated and the ease with which it can be controlled.
6. Pollution pathways are subject to a series of constraints.
7. We can detect pollution using chemical, biological or physical means. The most useful, basic methods involve biological indicators.

2.8.1   Introduction

Pollution is one of the byproducts of human existence and has probably been around for as long as we have grouped together and tried to use the environment. Today, we are more aware of our surroundings and so pollution becomes a major environmental issue. Because it's important and because we are polluting at a greater rate than at any other time in history it is essential that we get a grasp on the key concepts involved.

2.8.2   What is pollution?

This must seem like the most basic of questions and yet it's not straightforward. The problem is that any substance could, in theory, be a pollutant. Any definition must try to include all aspects without becoming unwieldy. There's no agreement about a standard definition. These are some of the answers:

'pollution may be defined as the introduction by man into the environment of substances or energy liable to cause hazards to health, harm to living resources and ecological systems, damage to structures or amenity, or interference with legitimate use of the environment.' (Holdgate, p.17)

'a pollutant is defined ... as a substance that occurs in the environment at least in part as a result of man's activities, and which has a deleterious effect on living organisms.' (Moriarty, p.3)

Examining these definitions raises three points. The first of these is that pollution is human-made but that for all pollutants there are natural counterparts. This means that all people do is to increase the range of items, not create whole new ways of disturbing the environment. For example, oil pollution is found naturally as oil seepage into the seas around Baja California in the USA. The second point is that pollution is not just a substance but also an amount. Very small amounts of an item can be found in the environment which do no identifiable harm. If we keep adding this item then there will be some small effect. At this stage we can call this item a **contaminant**. Beyond a certain level the effects become stronger and more noticeable – then we can call it a **pollutant**. One of the problems of pollution is that there are literally thousands of things that can become pollutants.

Tip

Remember the three basic rules – for every pollutant there's a natural equivalent, it is only pollution when it gets to a certain amount and just about anything can be a pollutant given the chance.

It is important to realise that pollution is not a twentieth century phenomenon. Some writers have argued that Neolithic people suffered chest disease from flint mining pollution. Certainly, polluted water supplies have been a major problem: lead pipes in Rome, plague in Medieval Europe and cholera right up to the twentieth century. In the UK there were attempts to control pollution going back to the thirteenth century when coal burning was a major source of air pollution around towns. The situation with air pollution was scarcely any better by the nineteenth century when Charles Dickens could write about the pea-soup fogs of London (in reality, smog). Although the situation was recognised there was very little that could be done. Virtually nothing was known of the nature of pollution and how it could be stopped. There were some early attempts to control coal burning but these just shifted the problem elsewhere. By the nineteenth century there was a greater level of understanding. Science was able to analyse pollutants and the growth of health biology (especially the discovery of bacteria) led to us understanding the causes and effects of pollution more clearly. This also led to the realisation that most pollutants could be dealt with. From this time onwards there has been an improvement in dealing with pollution and although the situation is far from over there's a far better understanding of the processes.

2.8.3 What is the nature of pollution?

Nobody doubts that pollution is having an effect but these effects are not evenly distributed. This means that there is something affecting the impact of pollution. There are three key aspects here: **source**, **pathway** and

target. The way a pollutant reacts in these three areas is crucial to our understanding. For sake of completeness, a pollutant starts with a source, travels along a series of ecological routes (collectively called pathways) until it reaches the final destination, a target.

A source is the origin of a pollutant: at this point it has just entered the environment. What is important is what it can do and how it moves through the environment starting with what the pollutant can do, i.e. its properties. Leaving aside the specific chemical and physical properties there are eight things that are important:

1 The **persistence** of the pollutant (the likelihood of it surviving in a state in which it can cause damage).

2 The **lifetime** of the pollutant (how long it's active for).

3 Interactions (chemical or physical reactions with other substances). Here, one needs to distinguish between the **primary pollutant** one that has an effect unaltered and the **secondary pollutant** which comes through interactions with other pollutants or the environment.

4 **Speciation** – the number of forms a pollutant can take. The greater the speciation the more ways it can react with the environment.

5 **Pollution loads** – the amount of a pollutant that can be found in the environment. Two concepts are important: **critical load** (the amount of a pollutant that is needed before one can detect damage to an ecosystem) and **target load** – the amount of a pollutant that is permitted or tolerable. Target loads would change with changing knowledge or even public perception.

6 **Dispersion** – the ease with which a pollutant would spread throughout the environment.

7 The ability to **bioaccumulate** – the ease with which a pollutant will stay in a target or environmental area and therefore build up.

8 **Ease of control** – how simple will the pollution problem be to clean up? A simple solid spill will be easy to remove, a liquid less so and gases almost impossible to contain. There are also hidden issues here. For example, when the first oil tanker disaster occurred in 1967 (the Torrey Canyon running aground off the Isles of Scilly) it was thought that the spill would be easy to contain. Unfortunately, it didn't burn and attempts to clean off oil with detergents created more problems than the oil did.

To make study easier people attempt to classify pollutants according to various criteria. The basic idea is that by grouping pollutants one can start to examine common properties and effects and even suggest some common strategy for solving pollution problems. Several systems can be put forward (see Table 2.8.1 for examples) although most writers use the **sector model** (e.g. air, water and land pollution) because of the ease for analysis.

Once the pollutant has left the source it starts to move through and interact with the environment. Although there are many ways of describing

Table 2.8.1   Classification of pollutants

1. **The nature of the pollutant:**
   a) Chemical composition e.g. inorganic/organic
   b) Physical state e.g. solid, liquid, gas

2. **Its properties:**
   e.g. solubility, biodegradability etc.

3. **Sectors of the environment:**
   a) air
   b) fresh-water
   c) marine water
   d) land

4. **Source:**
   a) fuel combustion
   b) industrial origin
   c) domestic origin
   d) agricultural origin
   e) military origin
   f) microbial activity

5. **Patterns of use:**
   a) industry
   b) home or in hospitals, schools, hotels etc.
   c) agriculture
   d) transport
   e) defence

6. **Targets and effects:**
   a) affecting the atmosphere
   b) affecting water processes
   c) affecting people directly
   d) affecting livestock
   e) affecting other animals
   f) affecting commercial crops and plants
   g) affecting ecosystems

such movements one of the most common is the use of a pathway diagram. A single pathway is a particular route from the source to the target. It's much like a road map. In modern pollution studies the pollution 'map' is drawn up to see where the greatest danger is. For example, when studying the impact of nuclear pollution from Sellafield in Cumbria, UK, it was found that fishermen who regularly ate local fish would be most affected (and that negligibly). The pathway from the fish to fishermen would be called the **critical pathway** – the route of greatest effect.

What does a pathway diagram look like? Study Figure 2.8.1 which shows in some detail the controls on a pollutant as it goes from source to target. Generally, as the pollutant enters the environment it will try to disperse. How and to what extent this is possible depends on a number of variables such as its rate of dispersion, its own chemical-physical properties (together, usually called the pathway-determining properties of the pollutant) and the properties of what it is moving in (e.g. air, water).

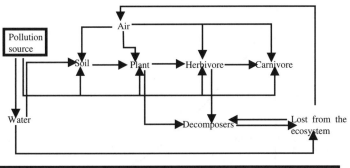

Figure 2.8.1 *Pollution pathways in an ecosystem*

Throughout the pathways there are places where it is removed (called pollution sinks), enhanced (e.g. bioaccumulation) or modified (by feedback loops e.g. eutrophication). The pathway is not just the environment; it also includes the target. The target has a series of pathways such as bloodstream whose absorption rate varies with pollutant and species.

This might look like there is a steady stream of pollution through the environment. Although this is true to an extent there are some other aspects to study. For example, what happens if the pollutant is released in bursts (called **episodic pollution**)? Acid rain pollution, for example, peaks during the spring melting of ice in Scandinavia leading to extreme levels in very short times. Another issue comes from the dispersion characteristics of the pollutant. Some might travel only a short distance. Our knowledge of meteorology (⇒ atmosphere) shows that most air pollution can travel long distances and cross political boundaries (**transboundary pollution**) which often raises the issue of who's going to pay.

Finally, the pollutant reaches the target. Concern here focuses on those factors which influence the final outcome: pollutant variables, ecosystem variables and target variables:

1 **Pollutant variables**. No two pollutants have the same effect. They will vary in **toxicity**, the poisoning power of the substance usually measured against a standard – Ld50 and Lc50 (which stands for lethal dose 50% or lethal concentration 50%). By exposing a group of organisms to varying amounts of a pollution source we can find out how much it takes to affect the population under study. Ld50 is the dosage that will kill half the population in the time (usually 24 hours) and Lc50 is the concentration of the pollutant that will have the same effect. Pollutants will also have different **dose:effect** relationships. Figure 2.8.2 demonstrates that the longer the exposure the lower the concentration needs to be to achieve the same effect. In general terms we would distinguish between acute exposure (immediate effect of a high

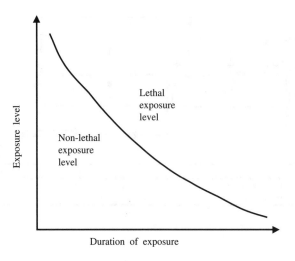

Figure 2.8.2 Exposure and effect

level of exposure) and chronic exposure (lower doses over a longer time).

2 **Ecosystem variables**. Ecosystems vary considerably in their responses to pollution: some will be damaged easily, others would appear to be more resilient: **stress** (ease with which alterations can be caused) and **resilience** (ability to withstand outside influences) are key concepts here. The **residence time**, the time the pollutant is active in the ecosystem, is crucial for many animals and has had a considerable effect in local systems. For example, after the explosion of the Chernobyl nuclear plant in 1986 radiation pollution spread through large areas of Northern Europe. This pollution had a high residence time in certain ecosystems such as the tundra in Scandinavia. This meant that the reindeer that grazed on the lichens that were soon too contaminated to eat and had a serious effect on the local economy.

Sometimes the effects of pollution will not be instant but may take years to show any impact (acid rain is thought to be like this). To explain this idea the cascade model has been proposed (see Figure 2.8.3). This means that individual parts of the ecosystem can withstand a certain amount of pollution before passing it on to other parts. Each drum represents a different part of the ecosystem's structure so that the initial impact will be biochemical: the last one will be complete ecosystem disruption.

Alongside the time lag effect between receiving a pollutant and passing it to another part of the system ecosystems can also alter the amount of pollution. Take the case of DDT, a pesticide with considerable side effects. Studies showed that it would be passed along the food chain but as it did so the amount in the receiving organism would

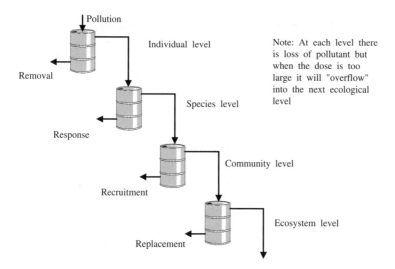

Figure 2.8.3 Cascade theory

increase. The amount of increase is shown in Figure 2.8.4. Because of this mechanism, what started out as a harmless amount became a lethal dose. This situation is generally referred to as **bioaccumulation, bioconcentration** or **biomagnification**.

3 **Target variables**. Studies have shown that larger organisms need more of a pollutant before the effects become obvious but there are also variations within populations. Some species (called **critical species**) are particularly susceptible to pollution. The canary in the coal mine is a simple example of this. Because it was much more affected by poisonous mine gases than the miners, if the canary was affected everyone knew to get out fast! This individual reaction can actually have a positive side. Consider the ability to withstand copper pollution. Tests on wild grasses in Anglesey led to the discovery that, on copper-polluted mine waste, not every plant would die. Over succeeding generations, a new copper-resistant strain of grass was produced.

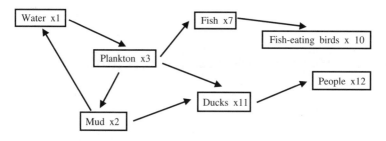

Figure 2.8.4 Bioaccumulation in a lake food web

2.8.4   How can we detect pollution?

One of the most important aspects of pollution study is the ability to monitor the environment, to detect pollutants and to assess their effects. Although there are numerous tests each with their own advantages and disadvantages it is possible to highlight some of the more common categories:

1 **Chemical**, e.g. biochemical oxygen demand (BOD). When dead organic material enters a river, micro-organisms start to break it down. To do this requires oxygen. BOD is a measure of the oxygen used i.e. how much oxygen would need to be added to bring oxygen levels back to the situation before the pollutant was added. Tests to assess this can be carried out using simple laboratory equipment (at least for an approximate result). Other simple tests can be used to detect pollution more directly. For example, a liquid pH (acidity) kit can be used to gauge levels of acid in rainfall or snow.

2 **Biological**, e.g. biological indicators. This group contains a wide range of ideas all centring upon the principle of the **bio-indicator species**. A bio-indicator species is a species with a known response to any given level of pollution. Thus the presence or absence of the species can give an approximate idea of how much pollution is present. The best-known example is a lichen experiment carried out by schoolchildren in the 1970s. *Plerococcus* (actually an alga) is the most resistant to sulphur dioxide pollution. This is followed by the 'crusty' group of lichens, the 'leafy' lichens and finally the 'shrubby' lichens which only live in clean air. Map the species and plot the pollution levels.

3 **Physical**, e.g. damage rating, cover diversity, Ringelmann scale. Physical methods rely on visual and general evidence rather than the individual species common in biological methods. Both damage rating and cover diversity scales could use the presence of numbers of species thus giving approximate results very quickly and cheaply (a major advantage of these methods). The Ringelmann scale uses graduated grey-scale filters to assess the density of smoke coming from chimneys. Although it doesn't show what's in the smoke it is an indication of overall damage.

2.8.5   How do we deal with pollution?

Before any solution is proposed there are a number of factors that need to be taken into account:

1 **Risk assessment**. There are four stages in deciding if a pollutant is dangerous enough to be considered a hazard for which control must be

sought: identification (what is the precise nature of the pollution), exposure assessment (how many people will receive how much pollution, using pathway analysis), hazard assessment (estimating the relationship between the amount of exposure and the risk to public health) and risk characterisation (the study of individual and population risk). What's interesting about this is that perception plays so large a part.

2 **Costs**. One of the biggest arguments in pollution control has been the issue of costs. Although there are a few exceptions virtually every pollutant has some method whereby it can be controlled. The difficulty comes when the costs are taken into account. For example, there is an old factory that will just about make a profit but which will not meet new pollution guidelines without extra equipment. The cost is greater than the profits will bear. What do you do?

3 **Applicability**. Many pollutants can be controlled by more than one solution; it's a question of finding the best place to control the pollution. For an individual factory a simple technological solution may be possible. However, the factory will also pollute the local area which will also need to consider some sort of response.

4 **What is the desired solution?** Some of the most recent ideas on pollution control look beyond the factory to consider purchasers and other stakeholders in the business. Thus the idea of putting in the first available piece of equipment may not actually be the best solution in the long term. For example, initial responses to organic pollution in rivers is to clear it up. However, if left alone the pollution will break down without the intervention of any chemicals. An early lesson was learnt from the 1967 Torrey Canyon tanker accident off the Scilly Isles. The oil reaching beaches was sprayed with detergent. This detergent killed nearly every species whereas the oil killed only a few!

Having considered these issues, appropriate techniques can be used. Most techniques fall into one of five categories. Firstly, there are economic measures. It might be tempting to go straight for this option because of the seeming simplicity of the idea – just charge people for polluting. However, there are some serious issues that need to be resolved. Where does a factory, for example, stop being responsible for its pollution? One of the more common approaches here is called the **polluter pays principle**. It is favoured by the EU and the OECD and puts the responsibility on the company to minimise pollution impact (or at least to keep within guidelines). One of the pioneering companies in this field – 3M – has tried another approach also going under the heading of 3P – **pollution prevention pays**. Their argument is that the cheapest way of dealing with pollution is not to produce it in the first place! The real difficulty facing economic solutions to pollution is that it is almost impossible to work out where to draw the boundary of responsibility and what to do once you've got there.

Secondly, governments can chose legal means. Companies are reluctant to reduce pollution if this means they pay more than competitors i.e. lose profit. One way of making sure this doesn't happen is to make a series of laws which say precisely what is allowed and what isn't. Most commonly, this comes in the form of a maximum permissible concentration for each pollutant (i.e. there's a limit to the amount of pollution that can be put in a place'. What happens when pollution might cause a problem but we don't know when/how much? Recent debate has brought in the **precautionary principle**. This means that we reduce pollution just to be on the safe side (it can't be removed after it has damaged the environment permanently). The banning of CFCs (refrigerants which might damage the ozone layer) is an example of this approach. As with economics, the legal side of pollution is becoming more difficult because our ideas of what's acceptable are changing. Some nations are considering **cradle-to-grave regulations** i.e. that in every stage of a product's life there are laws controlling how it's dealt with/disposed of. Recently, there's been discussions on following pollution in existing systems to clean up at every stage including before as well as after manufacture (**up-the-pipe regulations**). These examples deal with a single pollutant source. Some legal approaches have taken an entire region and sought to have legal control over pollution. The United Nations' Regional Seas Programme is an example of this.

Recent developments have provided us with a wide range of pollution control technologies. Since the range is enormous and increasing each day, some pollution technologists prefer to produce a solution that fits into a desired pollution state. Some of the more common examples of these pollution states include BATNEEC (best available technology not entailing excessive cost). The object is to control pollution within a series of financial limits and BACT (best available control technology) where solutions in this category will do almost everything wanted – at a price. These acronyms stand for an *approach* to using technology not a specific piece of equipment. Categories seem to centre around one key feature – cost. One argues for the best technology. It is recognised that the pollution can be dealt with; all that's needed is the correct system. The other camp argues for a solution that'll do as much as possible without costing too much. This might sound like a sensible way forward but how does one consider what costs are to be included? Where is the level for 'practicable' or 'available'?

Political solutions are best seen in the international setting. It is relatively easy to get pollution control in a national setting especially if there's public support. The 1956 Clean Air Act in the UK is a good example of this (following the London smog of 1952). With more pollution spreading around the world it is recognised that no one nation can deal with it. For this reason there are an increasing number of international groups trying to get a common agreement. Given the problems of long-range and trans-boundary pollution (especially where neighbouring countries are not politically friendly) there is a good deal of progress. The United Nations has a series of programmes in place in addition to the Regional Seas Programme. Other conferences have tackled sea dumping (the London Convention) and air pollution (the Montreal Convention).

Lastly, there are **multimedia** approaches to pollution technology. It has been recognised that using just one approach usually creates other pollution sources. The argument here is that all areas are checked simultaneously so that no new pollution sources are created. This entails sophisticated planning and design as well as technology.

2.8.6   Raising issues in pollution management

The final stage is to see what's happening in practice and what can be learned from our present approach i.e. how are we managing? Three cases will be used to illustrate this: acid rain, oil pollution and land waste management.

Acid rain was seen as one of the great pollution issues of the early 1980s but since then, public interest has died down in favour of global warming. However, it is still worth studying because of the lessons we can learn from it. To be accurate, all rain is acid and some acid rain isn't rain! The term acid rain (which has actually been known about for almost 100 years) is today used to link two phenomena: **acidified precipitation** (rain, snow, sleet for example) and **acid deposition** (dry acid particles falling to the ground). Figure 2.8.5 highlights both the sources and the transport of these compounds. When this material is deposited it has the ability to affect the environment although this depends on the conditions on the ground. Where the deposition is greater and the rocks and soil already acidic then there may be enough to create excess acidity in the surrounding ecosystems. This can leach through soils and into streams. Excess acidity can create several problems from leaf loss and die back in trees to death of fish and even health problems in small children.

What management problems can this case highlight? Firstly, there's the problem of finding out where the pollution comes from. Scandinavia suffers more than most from excess acidity. It wasn't until there were new discoveries in meteorology that the puzzle was solved. It was found that the link between upper atmosphere and lower atmosphere gave the perfect transport system for the acid rain. Pollution in Germany and the UK was taken up into the upper atmosphere to be deposited in Norway and Sweden (which explained why it wasn't found between these nations). This is the problem of science – to find out enough to say if there is a problem. Secondly, there's the problem of dealing with pollution control across borders. How can one nation control the pollution technology of another? Because there are no international laws on this there's little that can be done. Thirdly, there's the question of what to do even if one could get polluters to pay. One option would be to fit all pollution sources with easily

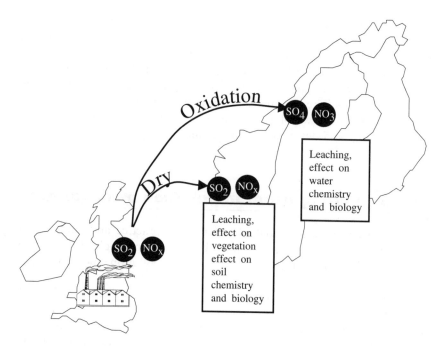

Figure 2.8.5 *Pollution: transport and effects*

available clean-up technology (such as air scrubbers to remove sulphur). The problem is cost. Some workers have calculated that only a small percentage is justified.

The second case is **oil pollution**. Oil is a complex series of hydrocarbons which, in crude state, is far removed from the oil and petrol we put in our cars. Pollution has a range of sources (see Table 2.8.2) of which ocean-based pollution is just one. Furthermore, oil can break down into a series of different products each with its own chemistry. The practical point of this is that one can't just set fire to it and solve the problem: older oil (say four days after a spill) has lost most of the chemicals that do burn leaving a tar ball that's difficult to cope with. This leads to a range of solutions from dispersants at sea to trapping oil (near to sensitive sites like nature reserves, for example) to physically scraping it off beaches. To add to this there are the ecological effects of these solutions. What management problems do we have? Firstly, there is the problem of finding the source. Tankers move on and rarely deposit near to land where they can be spotted. Small spills may not show up until too late. There are some techniques to 'fingerprint' crude oil by its trace elements but it's still a problem. Secondly, how does one respond to something that changes so quickly? It is possible to have a team standing by in every key location (but that still wouldn't have stopped most of the tanker accidents, only got to them quicker). However, is the cost-effective? How can an expensive crew and

Table 2.8.2 Estimated world input of petroleum by hydrocarbons to the sea

Source	Total
Transportation	0.555
Coastal installations	0.180
Other sources e.g. urban	1.380
Natural inputs	0.250
Production by marine phytoplankton	26,000
Atmosphere fall-out	100–4000

ships be justified for just a few days' effort? Thirdly, there is a need to examine the ways in which the problem is tackled. Early dispersants were far more dangerous than the oil. Sea birds dipped in detergent solution and released sank when their feathers could not hold air as the natural oil had been lost! Left alone, the tar balls would sink and degrade naturally.

Finally, the case of **land waste disposal** concentrating on domestic waste. Nobody can doubt that the amount of domestic waste we produce is increasing. Recent studies have shown that nations of the developed world produce between 300 and 900 kg per person per year (see Figure

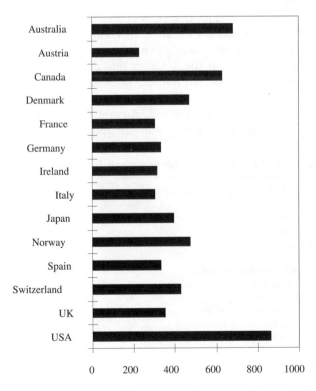

Figure 2.8.6 Domestic waste produced in the late 1980s per head of poulation

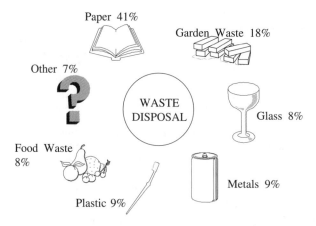

Figure 2.8.7 Waste materials

2.8.6) which is comprised mainly of six categories of waste (Figure 2.8.7). Much of this is, at the moment, put into landfill sites which is also the resting place for a considerable amount of industrial waste including toxic chemicals. The potential for pollution is high. For example, birds feeding on sites have their ecology altered and they can spread disease; water filtering through sites can pollute groundwater and some chemicals can damage the productivity of the soil for decades. In addition to these difficulties, many of the best landfill sites are full and there are very few people who want a new one opened near them. Given that this is a pollution source, what can be done to minimise the impact? Look again at Figure 2.8.7. The biggest waste category is paper. This could be recycled. So could other items. Many developed nations e.g. UK, Australia have adopted a similar approach to domestic waste which aims to reduce the problems. This

Table 2.8.3 Lifetime risks to life commonly faced by individuals

Cause of risk	Lifetime (70-year) risk per million
All cancers	196,000
Construction	42,700
Agriculture	42,000
Police killed in line of duty	15,400
Motor vehicle accidents (travelling)	14,000
Home accidents	7,700
Frequent airplane traveller	3,500
Pedestrian hit by motor vehicle	2,940
Background radiation at sea level	1,400
Tornado	42
Travelling 10 miles by bicycle	1
Travelling 30 miles by car	1
Travelling 1000 miles by jet plane (air crash)	1

involves organic waste composting, regular collections of recyclable material, using un-recyclable waste for combustion (to run a small generating station or to provide heat for a local area). This might seem like the ideal solution but there are difficulties. Firstly, any disposal causes pollution. If the waste is burned then it produces air pollution. Some items may be too insignificant to recycle (mobile phone batteries for example) but can produce toxic heavy metal air pollution. Secondly, recycling will also affect the economy of the primary plant (i.e. producer of un-recycled paper). Thirdly, it illustrates the problems of collecting from numerous sources. What are the economics of individual refuse recycling? Does it cost more or less in pollution terms for a car to be driven to a recycling bank with a few papers than for the papers to be thrown away? Would the extra personnel costs stop refuse collectors also collecting recyclables?

Questions

1. Find out about the domestic waste facilities in your area. What recycling takes place? Can these solutions be justified in terms of environment and economics?
2. Read through the definitions of pollution. Can you think of any pollutants that aren't covered by these definitions? How has this been possible?
3. Write down the percentage risk that you think each of the following would cause in a human lifetime (this figure would be the likelihood of being affected by the risk): cigarette smoking, agricultural work, drinking water contaminated with chloroform, living near a waste disposal plant, living near a nuclear reactor and eating peanut butter. Check with Table 2.8.3. What's the difference? Why did you think you got such figures?

2.8.7 References

Clark, R.B. (1992) *Marine Pollution*. 3rd edition. Oxford University Press.

Elsom, D. (1987) *Atmospheric Pollution*. Basil Blackwell.

Haslam, S. (1990) *River Pollution*. Belhaven Press.

Holdgate, M.W. (1979) *A Perspective of Environmental Pollution*. Cambridge University Press.

Holmes, G., Singh, B.R. and Theodore, L. (1993) *Handbook of Environmental Management and Technology*. Wiley.

Markham, A. (1994) *A Brief History of Pollution*. Earthscan.

Moriarty, F. (1988) *Ecotoxicology*. 2nd edition. Academic Press.

Tolba, M. et al. (eds) (1992) *The World Environment 1972–1992*. Chapman and Hall.

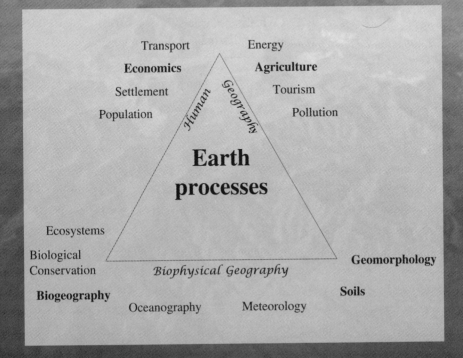

Transport   Energy

Economics   **Agriculture**

Settlement   Tourism

Population   Pollution

Human Geography

Earth processes

Ecosystems

Biological Conservation

Biogeography

Biophysical Geography

Geomorphology

Soils

Oceanography   Meteorology

Subject overview: the Earth and its processes

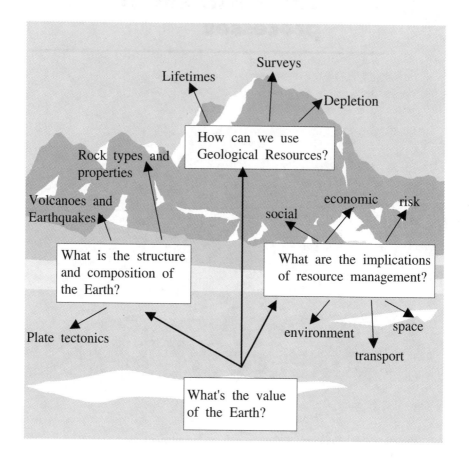

1. The value of studying Earth structure and processes comes from the value we place on mineral resources.
2. There are ten key elements in the Earth's composition. Most resources used by people tend to be present in far smaller quantities with implications for mining etc.
3. Plate tectonics is a unifying theory which can help to explain both the structure of the Earth and certain key processes within it. Since mineral distribution is often tied to such processes it has implications for resource location.
4. Volcanoes and earthquakes are examples of weaknesses at plate margins. Folds and faults are examples of weakness (competence) in rock.
5. Landscape formations depend on rock type, physical processes and time.
6. The resource base of a nation is a measure of its stability as a production area: resource lifetime a measure of an individual mineral's usefulness.
7. Resource management covers four aspects: exploration, exploitation, decision making and the impact of usage.
8. Implications of resource usage involve six areas of study: social, economic, risk, space, transport and environmental aspects.

3.1.1 Introduction

This chapter is about the study of the Earth and its processes. It is concerned with the structure and formation of the Earth and the resources we can get from it. As the demand for minerals increases so does the value of knowing what resources are where. This involves everything from formation to extraction. Increasingly this involves concern for the environment during and after extraction as well.

3.1.2 What's the value of the Earth?

As with ocean resources, as demand grows so does the importance of each mineral resource to the nation that 'owns' it. You only have to study the story of oil in the twentieth century to appreciate this. Such minerals give a nation three benefits. Firstly, there's the strategic value of having a large quantity of a resource within one's own sovereignty. If this resource becomes valuable e.g. uranium, then the nation with the most will have obvious advantages over other nations. Secondly, there are the economic benefits (although there are also disbenefits). For example, copper and diamonds gave Zambia and South Africa, respectively, considerable influence. On the negative side, such nations are unable to set the price; any day it could fall. This overdependence on primary production of minerals ($\Rightarrow$ economics) can create financial instability for a nation. Thirdly, there are the societal benefits of being able to use the resource. If your nation owns it, it can be used for the exclusive benefit of that nation. Many towns and even nations have thrived on having resource usage e.g. mining as part of the economic diversity of the area. This makes the study of the Earth one of the most important parts of geographical study.

3.1.3 What is the structure and composition of the Earth?

To understand the distribution of resources and to appreciate the features of the Earth it is important to know about its composition, structure and functioning. Whilst the composition, for our purposes, can be stated simply the structure is a more complex matter involving some of the largest forces on the planet.

Table 3.1.1 shows the basic composition of the Earth's minerals in both the crust as a whole and the mantle (an area just below the crust). Note that there is a distinct difference between the two sets of figures. The reason for this is that the Earth does not have an even distribution of minerals: iron and magnesium are more abundant towards the centre whilst silicon and oxygen are abundant in the crust. A cross-section of the Earth illustrating these points can be seen in Figure 3.1.1. The crust could be thought of as a thin, chilled layer which is constantly being fractured and pushed about by the convection forces in the mantle and asthenosphere. These forces, called **plate tectonics**, have proven to be one of the most important geological stories to have emerged in the last 30 years. When this question was first discussed in the early twentieth century the

Table 3.1.1 Composition of the Earth's crust

Element	% Weight
Oxygen	47
Silicon	29
Aluminium	8
Iron	5
Calcium	4
Sodium	2
Potassium	2
Magnesium	2

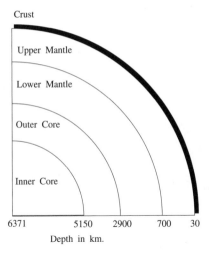

Crust
Upper Mantle
Lower Mantle
Outer Core
Inner Core

6371   5150   2900   700   30

Depth in km.

Figure 3.1.1   Earth's cross-section

main idea put forward was the 'wandering continents' hypothesis by Alfred Wegener. He proposed that continents 'floated' on the ocean crust and 'wandered' over the Earth's surface. Increasing evidence in the 1960s led to the proposal of plate tectonic theory. According to this theory, the Earth's surface is made up of a number of interlocking plates moved by underlying convection currents in the mantle (see Figure 3.1.2). There are three types of boundaries (Figure 3.1.3), each with a specific effect on the landscape. Constructive boundaries (where the plates are added to) occur mainly in the mid-ocean ridges and can have volcanoes associated with them (such as Iceland). The destructive boundaries have more dramatic effects because they are often associated with earthquake activity caused by the two plates creating friction as they are pushed past one another. The conservative boundary is one where nothing is added or removed (you'd need this for the Earth to keep the same size).

Other Earth movements can give rise to changes in the crustal structure (but not its location). **Volcanoes** are one of the more spectacular features. The giant plumes of ash and gas and the sudden destruction they can cause are serious reminders of the Earth's power. As such they are certainly hazards but they are also important in other respects. They are formed in one of two places. Perhaps the more important of these is associated with constructive plate margins e.g. mid-ocean ridges. Here, the weakness in the crust would be enough to allow lava to pour through and possibly reach the surface (e.g. Surtsey, near Iceland). In other cases there may well be a local 'hot spot' coinciding with a weakness in the crust. However it is formed a volcano is a hole in the ground (vent) through which molten rock (lava) and other materials can flow. What happens after that depends on a number of factors such as the type of lava, any other material present, the surrounding area, the force of the eruption and the

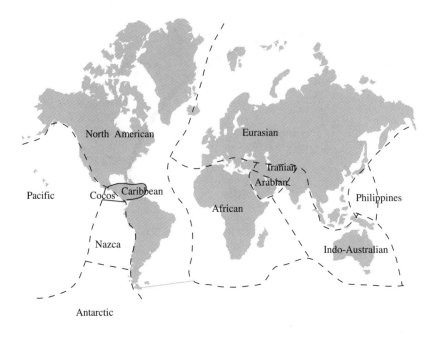

North American

Eurasian

Pacific

Cocos

Caribbean

Iranian

Arabian

Philippines

African

Nazca

Indo-Australian

Antarctic

Note: some smaller plates are omitted.

Figure 3.1.2 Tectonic plates

time since the last one. Volcanoes with regular fluid lava present a different landscape from those with infrequent, violent explosions. The contrast can be seen in Hawaii and Krakatoa. Hawaii has a fluid lava which is constantly being produced. Eruptions are common but cause little environmental disturbance. The 1883 Krakatoan explosive eruption on the other hand destroyed nearly half the island.

Closely linked to the volcano in terms of location is the **earthquake**. These are caused when friction between two tectonic plates at a destructive margin is sufficient to set off rock movements. The re-alignment of the rocks is the earthquake. Despite the majority of earthquakes forming in

Table 3.1.2 Richter scale

The Richter Scale is based on the relative intensity of the shockwave in the area it affects. Approximate values are:

4.2 – Slight – may be felt by people
4.9 – Slightly strong – wakens people
5.6 – Strong – objects fall off shelves
6.4 – Very Strong – walls crack
6.9 – Ruinous – ground cracks
7.3 – Disastrous – many buildings destroyed
8.2 – Catastrophic – total destruction.

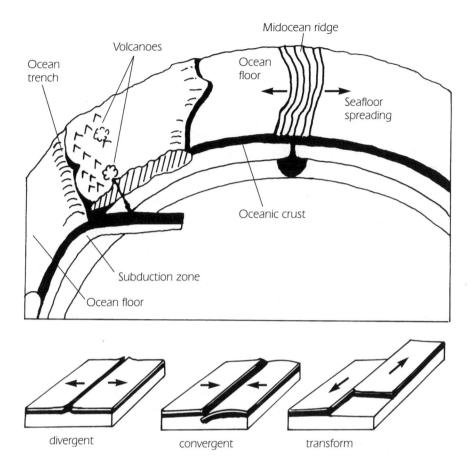

Figure 3.1.3 Three types of boundaries

this way there are other features that give rise to them although they would tend to be less violent (e.g. faults in fractured rock beds such as coalfields where small 'quakes are more common). The impact that these features have depends on how deep the earthquake was formed, how powerful the movement and how widespread the effect. Table 3.1.2 gives the common measure of earthquake power – the Richter scale.

Two other tectonic features take place over longer time periods. These are folding and faulting. When pressure is applied to rock layers (strata) they will usually fold: this is the basic process behind mountain formation. If the pressure continues or increases then the intensity of folding will also increase and so give rise to larger mountain chains such as the Himalayas and Alps. Sudden extreme pressure causes the rock strata to fracture giving rise to faults. Although such features are extremely important for geologists their key value to geomorphology is as a potential constraint on the development of rivers etc.

These processes produce the distributions of rocks and minerals. How do these processes give rise to features which can affect the way in which the landscape is formed? There are four main factors here: hardness, composition, competence and time. The **hardness** of a rock is a measure of its ability to resist being broken down. We compare hardness by using Moh's scale – a simple scale where any mineral is matched against a reference series. The key point is 7 – quartz (sand). Below 7 you have the softer minerals which usually mean rocks made of them can be easily worn down. Above 7 and the rocks become resistant to erosion (take Dartmoor for example with its uplands composed of granite – a quartz-rich rock). Allied to this is **composition**. Mineral hardness is only part of the picture. It also depends on how the rock was formed. Rocks are divided into three main groups: igneous, metamorphic and sedimentary which have a wide range of resistances to weathering. **Igneous rocks** form from molten material and produce rocks composed of crystals. The basic division in igneous rocks is between the granites and basalts. Granites are very resistant to decay because they contain large amounts of quartz (silicon dioxide as a chemical and sand as a particle). Therefore granites form the uplands in the UK such as Dartmoor. The chemical composition of granite is roughly the same as the continents themselves. The basalts form a more diverse group of rocks: they contain far less quartz and are more easily eroded. They are also far more widespread than granites. Basalts form the main part of the Earth's crust (i.e. the ocean floors) and are often seen on land as volcanoes (where the decay of the rock makes a particularly fertile soil). **Sedimentary rocks** are formed, as their name suggests, from sediments (or particles). These are the most common rocks on the surface of the Earth. They tend to be grouped according to their composition and particle size. The most important property from our point of view is the ease with which they decay. There's no hard and fast rule here because it can change according to rock group but generally the more the rock is lithified (turned into a rock by pressure or by being cemented by other chemicals) the more resistant it is. Major groups include the gravels, sandstones, clays, limestones (and chalk) and peat/coals. **Metamorphic rocks** are formed from an existing rock (of any type) through the action of heat, pressure and other chemicals. Because of their very diverse range there are few rules that can be applied to their properties. Generally, the more the rock has been transformed the harder it is but after that there's little to be said. It's worth noting that because of their minerals, metamorphic rocks are often sources of economic activity. For example, Dartmoor was surrounded by tin and silver mines in the nineteenth century. These metals were part of the stock added when fringes of the area were metamorphosed following the intrusion of a granite mass.

The third property is the rock's characteristics in response to stress: **shear** and **slide**. Imagine applying pressure to a rock surface. There are only a few things that can happen. It can remain untouched. If the pressure is great enough (or sudden enough) it might break (shear) the rock

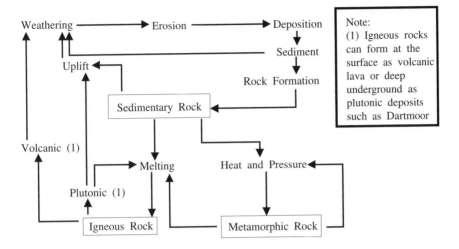

Figure 3.1.4 Rock transformations

and a sudden 'failure' can occur. Alternatively the stress can be small and/or longer term and the rock can adjust; this is called plastic deformation (slide). A rather useful term is **competence**. It refers to the way in which a rock reacts to pressure. An incompetent rock will slide or shear easily whilst a competent rock won't. This term is often used by civil engineers who need to know how a rock will take the pressure of a new road etc.

Finally, there is **time**. The longer the timespan and the larger the area and/or the greater the impact of the process. Allied to this is the main geological cycle: **rock transformation** – the idea that rocks are subject to numerous changes throughout time (Figure 3.1.4). This is closely linked, of course, with the processes at the surface ($\Rightarrow$ geomorphology).

3.1.4 How can we use geological resources?

Natural resource management is all about our use of the Earth's geological resources. Resources are generally taken to be physical items – minerals, rocks, land etc. In practical terms a resource is anything we need or use. The resource base is the sum total of resources available to a nation or area. Taken this way, a nation that had to import much of its raw materials (e.g. Japan) could be said to have a poor **resource base** (it doesn't have many resources within its borders). The USA or Australia would have a strong resource base because of the size of the country and the variety of rocks, minerals etc. within its boundaries.

Mention was made of time in relation to resources. Basically, how long will a resource last? Checking back over the past 30 years of North Sea oil reserve figures one finds that the reserves left have been about 30–40 years for each time they have been measured! Does this mean that all estimates are wrong? That's unlikely, what it means is that our idea of available resources has changed. The key idea here is that of **resource lifetime**. This is the period between discovery and depletion (there's a link here with the idea of the product cycle). For example, oil was known for centuries in the USA before it was used as a mass fuel. Similarly, there are still large coal reserves under the UK but it is not thought to be economically viable to get them. What affects resource lifetimes? Look at Figure 3.1.5 which outlines the key factors. Imagine this diagram applied to, say, uranium. The mineral is available and the technology in existence but the external economic and political factors alongside the use of substitute energy forms has meant that very little is now used. This means that because it's so little in demand its lifetime has increased considerably!

How do we calculate lifetimes? There are three basic models. The **Hubbert model** was named after M. King Hubbert of the US Geological Survey. He assumed that resource use would follow the maths of a normal curve where the area underneath the curve represented the total reserves. Subtract what you've used and you've got the total remaining. Such ideas might seem unlikely given the distribution of resources but they were an accurate reflection of US exploration/usage at the time. **Direct geological analogy** assumes that if x resources in were found in y area you get 2x resources in 2y area. Simple but an effective way of determining resources given a sufficiently good set of geological data. Finally there is the **Zapp hypothesis** – named after Arthur Zapp, a US scientist. He argued that after initial surface finds the rate of discovery would be a constant for every unit of exploration. Using oil exploration data he plotted the early finds and saw that later rates of discovery produced a nearly flat line (i.e. constant rate). This led him and others to argue that if you drilled in enough places then you'd find oil for centuries. During the 1970s this was an attractive idea especially because it meant we didn't have to worry about resources running out (just drill more). This is less popular today!

3.1.5   How do we manage our resources?

The management of resources can be divided into two parts: the obtaining of resources (exploration, exploitation and decision making) and the issues raised by using them (see next section). To illustrate this there are cases from two major resource areas: oil and mining.

The first thing one needs to do with most resources is to find them (i.e. exploration). Although each resource has a different technique the basics can be seen in the case of oil exploration. Assume we are trying to locate fresh oil fields to exploit. How would we go about it? One answer might be to send in a drilling team. This was a great idea in Texas with early oilfields

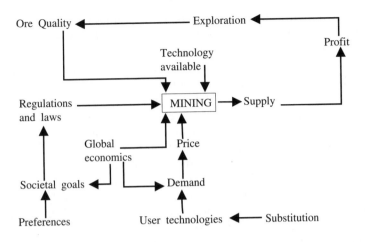

Figure 3.1.5   Factors influencing resource lifetimes

so close to the surface but of no use when dealing with the North Sea or current developments in NW Australia and Sabah. The ideal situation is to use a range of techniques each with their own costs and benefits. Remote sensing is the name given to techniques involving data gathering without direct human contact. The best example today would be the satellite survey. Here you have a broad picture approach for minimal cost per square kilometre. This can select sites for closer inspection e.g. pick out likely geological formations. Aerial survey, using planes equipped with instruments to detect magnetic changes in rocks, can refine the survey. Once this has been done, ground surveys can take place culminating in drilling if the tests recommend it.

Exploitation depends upon the nature of the resource and the demand for it. If the benefits of extracting the resource are greater than leaving it in the ground to appreciate in value (because of scarcity) then it will be exploited. This level of use is called the **optimal depletion rate** because it gives the greatest benefit. However, if we know the value that society can put on a resource in the future then we can judge whether we will get more benefit by using it now or later. Of course, we can't know the future with certainty and often we ignore the true costs of present production (e.g. ignore pollution and disposal costs). The outcome of using this idea is that we will tend to favour short-term usage over long-term conservation. This means that any resource management has to overcome a considerable body of established ideas (which is one reason why renewable energy resources have made so little headway over fossil and nuclear options until recently).

If resource management were just a case of exploration and perfect market economics then we'd have a less difficult time making sense of it all. In the real world there are other factors involved and this goes under the heading of decision-making. Because of the importance of resources to nations then there's always going to be some politics involved ($\Rightarrow$ politics). The best example here is the 1973 'oil crisis'. Oil-producing nations formed a cartel (a group of people acting together, in this case, OPEC) to create a price structure favourable to them which raised oil prices four-fold. A key idea was to promote national control of resources (rather than let oil companies do it) although resource conservation was also mentioned. Other groups, called stakeholders, can also be involved (e.g. shareholders, consumer groups, environmental groups etc.)

Mining provides a second case. In terms of resource management there are three ideas worth examining here: proven reserves, usage patterns and strategic planning. **Proven reserves** are those parts of the resource whose location and extent are known. The reserves, normally measured in years, vary greatly depending upon a range of factors. This means that we can get figures ranging from 18 years estimate for silver to 400 years for diatomite (a clay) whilst other reserves e.g. silicon are far beyond this. The implication here is that there is not just one date but a whole range and therefore management must concentrate on each individual resource and not a group. **Usage patterns** are also part of the equation. For example, what degree of substitution is possible? For sulphur there are no real substitutes but for copper there are numerous metals and plastics that can do the same job. This means that it's not just reserve policy that alters life-times but also the way in which the resource is consumed (or not). Within this idea comes the notion of domestic supplies and import reliance. Domestic supplies are those reserves under the direct political control of one nation whilst import reliance is the measure of how much nations are dependent upon others. If the resource is of little significance then there is no real problem. The issue becomes important if the resource is a direct and vital use to the nation. Finally, there is the idea of **strategic minerals** and stockpiling. Strategic minerals are not just those related to oil and metals; gravel and limestone can also be seen as strategic in the UK. The value is not the price of the resource but the importance of what it can do. Thus we cannot say that gravel is important in cash terms but it is vital to building! One practical use of this idea is in mining in National Parks in the UK. Whilst this is not allowed in general, if it can be proven to be of strategic importance then mining will be allowed despite the conservation measures (called the Silkin doctrine after the politician who first put it into practice). Stockpiling can be seen in both economic and political terms. Economic planning requires that there are sufficient supplies to keep factories going (e.g. coal stocks at power stations) whilst on the political front, a nation may want to stockpile resources to ensure a safe supply (or even to push up prices).

3.1.6 What are the implications of resource management?

What of the management issues in mining resources? Given that mining is far more than just the pit or well and is often closely linked to the surrounding community there are three aspects – social, environmental and political, that need to be taken into account. On the social side, both expansion and closure have impacts. Closure of a mine often brings widespread unemployment with corresponding loss to the whole community. Following pit closures, coal-mining areas have been devastated as communities and even after 20+ years there are still few signs of total recovery in places like South Wales. Interestingly, whilst there are controls for the environmental and economic impacts of resource changes there are virtually none for the social side. Environmental concerns have become increasingly important in the last few decades. Stricter controls have reduced pollution at the mine during use and although there is a great deal to do, there are definite signs of improvement. Table 3.1.3 shows the range of problems that can affect coal mining. One of the newer environmental concerns involves recycling. There is no doubt that recycling is gaining in importance and will continue to do so. However, for resource managers there are still questions that need to be asked. For example, if the cost of recycling is more than the raw materials how do you justify recycling? As recycling

Table 3.1.3 Environmental effects of coal mining

Air:
    Oxides of sulphur, nitrogen and carbon
    Particulate matter
    Radioactive material

Water:
    Acid water
    Mine liquid
    Waste disposal
    Coal washings water
    Hot water (from power stations)

Land:
    Loss of land for building
    Subsidence
    Loss of habitat
    Loss of historical features

Wildlife:
    Habitat disturbed
    Wildness lost
    Exploitation of area

Miscellaneous:
    Noise
    Dust
    Visual impact

is still in its infancy one can be sure that other questions striking at the centre of this issue will become more frequent.

Although each case of resource usage is unique there are six sets of implications which could be seen as common to all of them:

1 **Social implications** can be seen in terms of both politics and equity. In political terms all cases involve laws and regulations – it is accepted that resource management could not be carried out without some control. In turn this raises the issue of jurisdiction – who is actually in control? This might be easy if the whole matter stayed within one nation but the more global issues e.g. ocean usage and pollution create a new set of difficulties. Equity is becoming a more important term. In sustainable development (see below) it is seen as crucial. Who owns the resources and what rights do future generations have (called intergenerational equity)? How are we to distribute key, scarce resources? Given our increasing pressure on resources such issues will not go away.

2 **Economic implications** are seen in resource management. We need to be able to compare present value with future cost to find a solution to the rate of resource consumption. Reserve estimation is a difficult geological calculation but it is also a financially difficult one. The rise and fall of prices in the world's stock markets and the closure of coal mines with reserves still workable highlight the issues. Environment and economics are becoming more linked and recycling, substitution and technology are all changing the picture. Environmental concern is also showing itself in business with most large firms having some sort of environment programme in operation.

3 **Risk implications** are also seen. Some minerals e.g. asbestos are hazardous and create their own risks in terms of mining and usage. Risks are also seen in stockpiling where national security must also be balanced against economic issues.

4 **Space utilisation** might be important for the high-density population of the UK but also affects industry and transport. For example, oil refineries must have sufficient space for safety and airports leave vast areas untouched to that they can operate with minimum danger. If we are concerned about transport then we might examine the use-mix of the land. This is the idea that you have a variety of land uses so that what the individual needs is never too far away. The implication is that we should try to reduce our impact however and wherever possible.

5 **Transportation** is a major issue in geography not just for the movement of goods and people but also for its environmental impact. In terms of resources it is a major consumer and so we need to examine it carefully. With infrastructure issues we must also consider our need for transport. What is the correct balance of public versus private? Do we need to travel to work or does the work come to the home? Such questions are just being recognised but they will become more important as we accept that transport has its limits just like any other resource. One

study in the USA in the 1970s assumed that as fuel stocks fell so the last fuel would go to airplanes. Would we feel the same way today? What is the correct mix between speed (e.g. Concorde) and bulk carriage?

6 **Environmental implications** have been seen throughout this chapter. Several issues are seen as being common. Firstly, we now accept the interdependence of the Earth's resources. If everything is linked in some way then it means that resource management must take this into account when planning resource usage. In ecological terms there are areas of considerable fragility and areas easily stressed. This suggests that we need to take care in our use of resources and plan to reduce impact wherever possible.

___ **Questions** _____

1. Take a resource local to your area (like sand or gravel, it doesn't have to be big). Find out about its management.
2. Choose a resource. What laws and policies govern its exploitation? Which stakeholders are involved and what is their role? How could you resolve any differences?

3.1.7 References

Brereton, E. (1992) *Resource Use and Management*. Cambridge University Press.

Cutter, S.L., Lambert Renwick, H. and Renwick, W.H. (1991) *Exploitation, Conservation, Preservation*, Second edition. Wiley and Sons.

Owens, S. and Owens, P.L. (1991) *Environment, Resources and Conservation*. Cambridge University Press.

Smith, L.G. (1993) *Impact Assessment and Sustainable Resource Management*. Longman Scientific and Technical.

3.2 Geomorphology

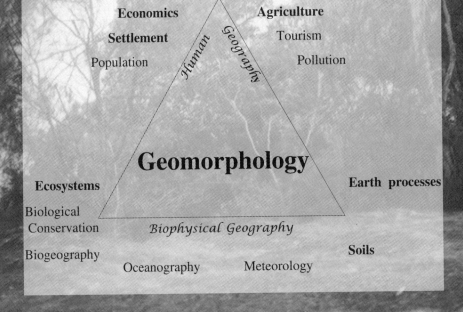

Transport   Energy

Economics   **Agriculture**

Settlement   Tourism

Population   Pollution

Human   Geography

Geomorphology

Ecosystems   **Earth processes**

Biological
Conservation   *Biophysical Geography*

Biogeography   **Soils**

Oceanography   Meteorology

Subject overview: geomorphology

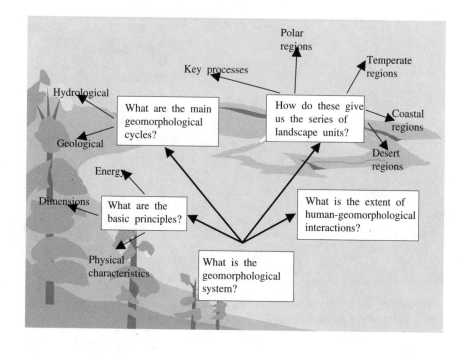

1. Geomorphology is the study of the changes to the surface of the earth. As such it is concerned with the action of atmospheric and hydrospheric processes on the land.
2. Key variables in understanding the effects of geomorphology are rock type, the amount of energy involved and the length of time the process has been acting.
3. The energy in any geomorphological process comes from one or more of convection currents, gravity and potential/kinetic energy (the energy of movement).
4. The geological cycle is based on the formation and degradation of the three main rock types: igneous, metamorphic and sedimentary.
5. Global-level processes are based on the plate tectonic theory and many key phenomena – volcanoes, earthquakes etc. are associated with plate patterning.
6. Weathering, the breakdown of rocks, has three main methods: physical, chemical and biological. All are active everywhere but some may be so slight as to make no appreciable contribution.
7. Erosion is the wearing away of rocks and involves the transportation of material from the site of the erosion. Thus common forms of erosion are wind and water based.
8. Slopes are a basic unit of study. How they form and how they control the landscape is a major feature.
9. Each major landscape unit has all processes at work in different amounts.

3.2.1 Introduction

Geomorphology is the study of the Earth's crust and the processes that shape it. Today, with human impact becoming greater it is important to know how the Earth's surface will respond to change e.g. the building of roads and airports. Perhaps the easiest way to start this study is by using the 'landscape unit'. This is an area, clearly recognisable, within which a series of processes shape the land. One of the best examples of this is the river basin (sometimes called the catchment area). By examining all processes that take place and knowing the factors, like geology, that control development it is possible to understand the resulting landscape.

3.2.2 What is the geomorphological system?

Physical geography often uses a systems approach in its work. At the largest scale there are four key parts: atmosphere ($\Rightarrow$ meteorology), hydrosphere (water), lithosphere (⇒ geology) and biosphere (⇒ biogeography). Their interrelations can be seen in Figure 3.2.1. These four aspects interact in any given place to produce a unique landscape. For example, despite its diversity in other matters, polar regions are easily recognised by the impact of ice.

3.2.3 What are the basic principles?

Much of the work on the four main 'spheres' has been mentioned in other chapters: here, the focus is on the influence of water. Since much of the temperate landscape of the world is dominated by water processes it is appropriate to concentrate upon it here. There are three main areas that need to be studied. The first of these involves the movement of energy. It is energy in its various forms that creates the forces used to alter landscapes. Much of this is shown by the work of the second area, the water or hydrologic cycle. Finally, there are the processes of weathering and erosion, the two elements that wear down the Earth's surface.

Energy is the force that drives all the others. **Gravity** acts as a constant downward pressure on slopes. If added to other forces it can explain a range of phenomena such as mudslides, avalanches and erosion. The key here is to consider the relative force of gravity pulling downslope and the resistance (inertia) of the rock on the slope. If the slope has less friction or the rock is less competent then it will move downslope. If the angle is too great then a rockslide could occur. Potential and kinetic energy, associated

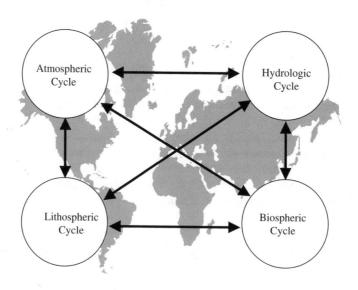

Figure 3.2.1 Interacting cycles

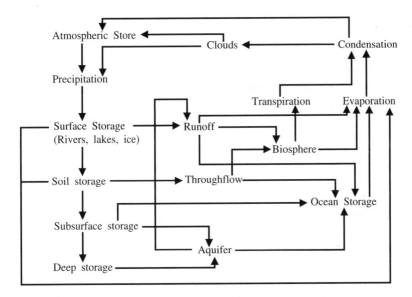

Figure 3.2.2 *Hydrological cycle as a systems model*

with movement, can be used to explain a range of phenomena such as erosion. Something has **potential energy** by virtue of its position. For example, if a ball is on top of a table then it has potential energy because it's above the lowest point (the floor). Once the ball starts to drop it loses potential and gains **kinetic energy**, the energy of motion. This 'release' of kinetic energy (giving the ball 'momentum') means that it might be possible for the rock to erode. This can be used to explain river erosion, deposition and transport. Lastly, there's the **least energy principle**. For example, rivers tend to erode to fit to an ideal line (grade) because it represents the least energy situation. Sometimes the actual line taken (profile) is above or below that line. By using these few basic ideas you can work out that any part above the line would have more potential (and less kinetic energy) at that point than you would expect. Thus erosion will be increased. However, if the profile is below the line then it has less energy than you might expect and so deposition could occur.

To understand this more fully it is necessary to study the water or **hydrological cycle** (Figure 3.2.2). Water falling on the system as rain or snow is intercepted by trees etc. Water drips down the stems and off leaves although some is evaporated back to the atmosphere (40% or more in summer). That which reaches the soil might run off (overland flow) or seep (infiltrate) into the soil where it either stays or flows through. Eventually it reaches deep water storage in rocks. The length of time this takes is of interest to geographers. One study of chalk streams in Southern England showed a **residence time** of over 20 years (when they thought it might be 2 or 3). The *global* water cycle is an expression of the amount of water held in the major parts of the planet (mainly the oceans which have about 97%

and the atmosphere which is only about 0.001%) and the way in which it moves around. Key processes are **evaporation** (which takes water from a liquid to gas state) and **condensation** (which does the reverse). Globally, of course, the system is in equilibrium but locally, there are significant differences. The **water budget** is an attempt to examine these differences. Water supply (through rainfall etc.) is plotted against the amount likely to be lost (from evaporation and use by plants – transpiration). Using this we can see whether or not a shortage is likely to occur. The **river basin**, a *local* part of the water cycle, acts as a single unit because within its boundary (called a watershed) all precipitation goes to one river system. For example, it is possible to say that an area like the UK is a series of river basins – Thames, Severn, Tees etc.

To appreciate the way the water cycle affects that landscape it is necessary to understand some of the features studied and see how they relate to our understanding. To start with it is necessary to have some measure of water movement and the easiest one to use is the **streamflow** i.e. the amount of water in the river. This can be measured by a gauge and produces a graph called a **hydrograph**. Figure 3.2.3 shows the basic terms and ideas. There are two things worth noting especially: **lag time** and **discharge rate**. The lag time is the time taken from the end of the rainfall to the peak flow and is a measure of the cascading capacity of the river basin – its ability to remove as soon as it has fallen. A short lag time would suggest a rapid movement of water through the system. The discharge rate gives more than just the quantity of water involved. The shape of the two limbs can tell us about the characteristics of the river basin.

Sometimes it's more useful to have an annual figure rather than the daily hydrograph. Farmers, for example, might use such a figure because it could tell them the likelihood of flooding (or more often the effects of drought and the usefulness of irrigation). This figure is called the **hydro-**

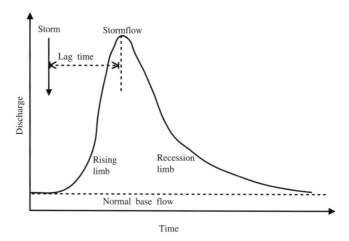

Figure 3.2.3 Storm hydrograph

logical equation: Streamflow = precipitation – potential evapotranspiration +/– changes in storage. (The potential evapotranspiration is a measure of the amount of water that should be lost through evaporation from the soil and loss through plants.) There are two other useful measures. The **discharge ratio** is the ratio of discharge as a percentage of the precipitation (D/P × 100%) which can show how much gets into the soil system. The average annual run-off gives the mean value for streamflow and is useful in comparing flood/drought readings. This is calculated by dividing the average annual discharge by the river basin area (also called the **catchment area**) i.e. D/A. As with any measurements there are going to be factors which cause variation. In relation to hydrographs these are rock porosity, human activity, relief and the drainage network.

This last variable is important to our understanding of catchments as it involves the size and shape of the basin. Four measurements are usually made. The first is **drainage density** – the total length of all the streams divided by the basin area. Secondly, there is **stream ordering**: the numbering of each stream according to its place in the system. Thus streams stretching from the source to the junction with another are first order. Two first order streams will join into a second order and so on (Figure 3.2.4). From this we get the bifurcation (or forking) ratio. It's the ratio of streams in order 1 divided by those in order 2, thus: BR = (no. of 1/no. of 2 + no. of 2/no. of 3 + no. of 3/no. of 4...) / no. of fractions. Given 20 first order, 15 second order, 5 third order and 1 fourth order stream the BR = (20/15 + 15/5 + 5/1) /3 = 3.1. The **circularity ratio** (CR) compares the area of the basin under study (AB) with the area of a circle of the same perimeter size (AC). Therefore: CR = AB/AC. Finally, the **elongation ratio** (ER) compares the diameter of the basin area as a circle (DB) with the maximum length of the basin (ML). Therefore: ER = DB/ML.

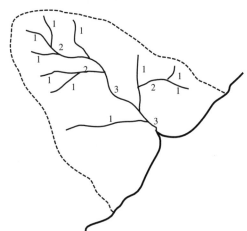

n.b. as each tributary joins another the order increases

Figure 3.2.4 *Stream ordering*

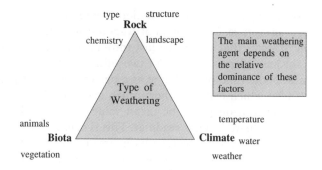

Figure 3.2.5   *Factors influencing weathering types*

Finally, there are the two processes that shape the landscape – weathering and erosion. **Weathering** is the breaking up of rock into rock particles. **Erosion** occurs when weathered particles are transported and therefore have enough kinetic energy to wear away other rock surfaces. If there isn't enough kinetic energy then the particle will be deposited (some writers put both terms together and call it **denudation**). In virtually all places there's weathering, erosion and deposition going on all at once: only the proportion of each changes as one goes from one ecosystem to another. Figure 3.2.5 shows the key factors that determine which form of weathering will be dominant - physical, chemical or biological.

Physical weathering relies on changes in temperature or pressure. This type of weathering is common in polar regions but can be found in any areas where climatic conditions are the key factor. If water can enter a crack in a rock and then freeze, the pressure it can produce might be enough to split the rock. This process, **freeze-thaw** or **frost shattering**, is common in the UK where a severe frost or long spell of wintry weather can break up road surfaces. In the landscape, prolonged freeze-thaw gives rise to a mass of angular rock fragments which, close to a cliff face, we call scree. **Exfoliation** operates in deserts and similar places. If a rock is heated up, the surface will get warmer (and expand) faster than the inside. Differences in heating cause tensions which can flake layers off the rock. Rocks form in balance with their surroundings. If a rock is formed deep underground then it will be able to withstand the pressures of that depth. Bring it to the surface faster than it can adjust to and tensions in the rock can cause it to split which is called **pressure release** weathering. Although this is a very minor form of weathering it does have serious implications. Tunnel builders in the Alps have known that, as they bore through the rocks, they release the pressure of the rocks in the roof and sides. It has been known for some slabs to become detached and fly down the tunnel causing injury.

Chemical weathering relies upon chemical reactions to break down rocks. **Solution** (also called carbonation) weathering is common in areas with chalk and limestone. A weak acid, usually carbonic acid in the rain, reacts with the limestone to form a soluble chemical – bicarbonate. If this continues for long enough then we get the landscapes characteristic of limestone regions such as caves. This process is best seen for most people in our cities where acid pollution has dissolved the statues around churches and civic buildings. Some rocks will break down if water can react with some of the minerals making it up (**hydrolysis** weathering). One of the best examples is granite. It's very resistant but it does have a weakness – one of the minerals can react with water to produce one of the clay minerals. With this mineral gone the others will be easily weathered. Alternatively, **hydration** causes some minerals to absorb water and expand. This can lead to the rock being broken up. Some rocks will contain minerals that can react with oxygen (**oxidation** weathering). One of the best examples is iron. If the amount of iron in a rock is high enough then when it reacts with oxygen (usually in water) it will change chemically – in simple terms it will rust! This might well lead to the break-up of the rock by other forces.

Biological weathering can be a factor in some areas. There are cases of animals breaking down the rock directly. One case is where elephants use their tusks to dig into a band of salt in the rock in East Africa. They get the salt they need and the area gets a new cave! Rabbits might not create such problems but they can continue to break down rock into smaller particles (most rock disintegration needs more than one process). Trees and tree roots can produce pressure similar to freeze-thaw action and can split rock by growing up through it (a common sight in urban areas with trees and even weeds breaking through tarmac surfaces).

The type of **erosion** that occurs depends on the area. In coastal areas **wave action** can break down rock by the physical power of waves. Alternatively, a wave may break so that it traps a pocket of air underneath it. This **pneumatic action** as it's called might be enough to force air into rock cracks and start to weaken it. River erosion is a major force in many areas of the world especially in temperate zones. Rivers can widen and deepen their beds using a variety of methods. For example, **corrasion** is a process where particles scour a surface and start to wear it down. In some places the sheer force of water will erode banks – a process called **hydraulic action**. With additional pressure and even bubbles caused by rapids, **cavitation**, an extreme form of hydraulic action, can occur. A river can also be considered a stream of chemicals from which one gets **chemical erosion** (e.g. limestone). Lastly, rocks in the water can be chipped and eroded (the process is called **attrition**). Whatever particular process, or set of processes, is operating there is one key diagram. Called the **Hjulstrom curve** it shows the relationship between velocity, erosion, transportation and deposition (Figure 3.2.6). Too little energy and particles will be deposited. If there's an 'excess' then erosion takes place.

Glacial erosion involves the effects of ice: **abrasion**, where the rock is scraped away by the passing of a debris-laden glacier and **entrainment** (where rocks are moved by being locked in ice and taken up by the glacier).

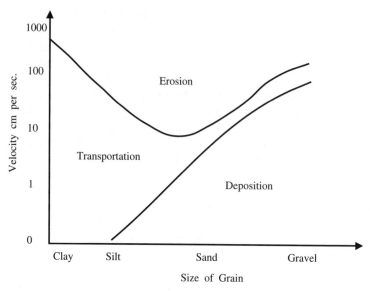

Figure 3.2.6 Hjulstrom's curve

Just as the Hjulstrom curve is crucial in understanding the actual processes in rivers so the numerous factors in glacial regions determine the precise process. Figure 3.2.7 summarises these factors. **Wind erosion** is a feature of arid regions such as deserts. In many ways the process is similar to water erosion. Particles may abrade an area (the natural equivalent of the industrial process sand-blasting). Alternatively, the wind might pick up grains and carry them over several kilometres leaving behind **deflation hollows**.

3.2.4 What landscapes can these processes produce?

It was G.K. Gilbert who first put slopes on the map. He argued that all landscapes were composed of slopes and therefore their study should be a key part of geomorphology. This idea has implications: that similar slopes are subject to the same processes and outcomes and by analysing slope patterns we can work out the physical history of the area. The **classification** of slopes is important because from that we can start to discuss the controls and processes that made the pattern. Figure 3.2.8 shows the four main slope elements: convex, free face, straight and concave. Note that with each one there is an associated erosional force: creep and rainsplash, mass movement and surface wash. Key slope controls include rock type, climate and geological structure because they provide the basics upon which the erosional processes are to act. One aspect worth further atten-

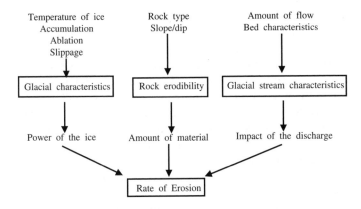

Figure 3.2.7 *Factors influencing glacial erosion*

tion is the feedback mechanism two key controls of which are vegetation and soil. With the development of both soil and vegetation you have an 'insulating' layer between the climate and the bare rock. Acting as a buffer, this must slow down the process.

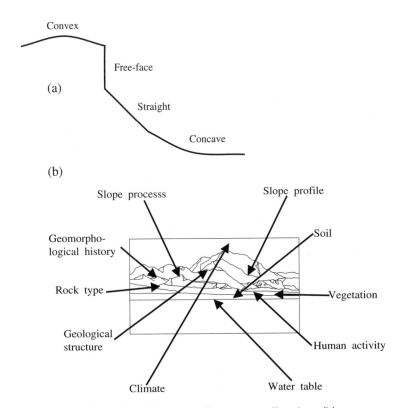

Figure 3.2.8 *Slope forms (a) and the factors controlling them (b)*

Although it's possible to write about a single process, in reality slopes are subject to a range of processes at the same time. **Slide** (sometimes called mass movement) is the transport of material through the action of gravity. A rock particle moves when the effect of gravity is greater than the frictional force keeping it in one place. From this we get landslides or rock-slides. When there is sufficient moisture then **flow** will occur. At the bottom of the slope it might be surface wash but with a steeper slope it could be a mudslide. Finally, there is **heave**. This is a slow process which involves the repeated freezing and thawing of a surface. Eventually, the surface will shear and give rise to the small ridges characteristic of slopes in chalk areas. One element not accounted for is **time**. If these processes act for long enough slopes should evolve a characteristic pattern. Three theories have tried to explain how this could happen. According to W.M. Davis the angle of slope is reduced (i.e. declines) by surface erosion (**slope decline theory**). For W. Penck the steep part of the slope gradually erodes back to be replaced with a gentler slope below (**slope replacement theory**) whilst for L.C. King the initial form of the slope is kept and the only change is the parallel movement of the slope faces as they move away from one place (**parallel retreat theory**). Given that examples of each slope type occur then it's possible that all three processes are in operation.

The next stage is to describe key landform units in turn. **Polar landscapes** are characterised by extremely low temperatures and minimal solar radiation which means that physical weathering and glacial erosion are the dominant forces. The climate can be divided into two: polar, where the mean temperature is below 0°C all year round, the ground is covered in ice and there is no plant life and tundra where mean temperatures are between 0°C and 10°C with some plant life. Despite the snow cover there is usually little precipitation in polar regions (some places in Antarctica are amongst the driest on Earth). Polar landscapes are controlled by a number of factors. The accumulation and compaction of snow is important. Once the temperatures are low enough to allow the permanent collection of snow, pressure of additional snow starts to compact the lower layers (called firn or neve). From here, we can examine the glacial budgets and systems. The glacier is like any other system with input (snow) and output (water, if the temperature is high enough). Once there is enough ice then the mass will start to move. In cold areas where ice accumulation is high the ice-sheet as it's called will be frozen to the surface and very little erosion will take place (which explains why change in the Antarctic is so slow). In more temperate areas only smaller areas such as valleys will be affected by ice – the glaciers common in the Alps, for example. The physical impact of the glacier is considerable. This results in several forms of erosion: abrasion, plucking, rotational movement and extending and compressing flow.

By contrast, **temperate areas** are often dominated by rivers with river erosion as the obvious key feature. Energy depends upon height above sea level. The long profile is a source-to-mouth cross-section of the river. In its ideal form (grade) it shows a balance between erosion and deposition.

Near the source, gravity is the main force and the stream will erode downwards giving small, steep-sided valleys. Further down the sideways movement will create wider channels. W. Davis argued that river valleys could be divided into 'youth', 'maturity' and 'old age' each with their own characteristics. Typical of the river landscape is Figure 3.2.9. The 'aim' of a river is to erode to its base level (i.e. the sea level). What happens if the sea level changes? In the case of a lowering of sea level, the potential energy of the river will increase and, at each point along the river, there will be more energy than previously. This would allow greater downwards erosion to occur in the process known as rejuvenation. Commonly, the river would leave a series of terraces either side (as seen in London for example). Sea level might also rise in which case there would be less energy. At this stage the river might develop several channels seemingly randomly in a braiding pattern. What of variations in the long profile which might be temporary features? Imagine the long profile with a layer of harder rock at one point. Rock either side would be worn away leaving the hard rock above the normal surface. Being above grade would imply that there would be more energy and therefore greater erosion. We would refer to such a point on the long profile as a nick point and the feature would be a waterfall.

Depending on how you define **deserts** it's estimated that up to one-third of the land surface is under arid or semi-arid conditions (i.e. desert or semi-desert) where arid can be defined as an excess of evaporation over precipitation. Meigs took this one stage further by calculating an aridity index from which he argued for three desert types: extremely arid, arid and semi-arid. Deserts can also be classified according to the geology of the area: mountain and basin desert (e.g. USA); plateaux deserts where layers of sandstone rock have been cut through by desert valleys (or wadis); shield deserts composed of eroded crystalline rocks e.g. Western Australia; sandy deserts or ergs such as the Sahara and stony deserts or regs. Alternatively, temperature can be the deciding criterion giving: hot deserts

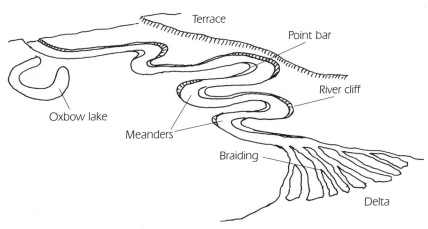

Figure 3.2.9 River landscape

such as the Sahara in North Africa and cold deserts such as the Gobi desert in Mongolia.

Today, we are concerned with the spread of deserts (desertification) because of climatic changes and human impact. Therefore, it's important for us to be able to understand how deserts can form. The main factor, aridity, is caused by a lack of rainfall which can be due to a variety of atmospheric phenomena. A high pressure area can 'push' moist air away from certain areas such as the Sahara. In some places, a rain shadow (i.e. in the lee of a mountain where there's very little rain) can be sufficiently strong. In coastal areas, coastal deserts can form because the cold sea currents parallel to the coast do not allow rain-bearing winds inland. Finally, there's the idea of continentality. In an area as large as a continent there'll be places too far for the sea to play a moderating effect. Such places tend to be hotter or colder than one might expect – rainfall could also be far lower.

What are the major desert landscape processes? As you might expect, water erosion is minimal with the exception of the rare, violent flash flood. Wind erosion can be significant especially where there is sufficient sand to create a sand-blasting effect. Given the daily temperature variations in places such as the Sahara, it's not surprising that exfoliation weathering is seen. Although there's little water there is a form of chemical weathering. Just as ice can split rock, so can the formation of salt crystals. If there is enough moisture then salt can rise by capillary action to the surface. Once there salt crystals can form in small cracks and start to widen them. With so little vegetation there's little to stop the erosion of desert areas. Landforms tend to reflect this. In terms of erosion features we can see isolated hills (e.g. buttes of Montana, USA) called **inselbergs** surrounded by their smooth surfaces or pediments. Sometimes the erosive effect can create strange shaped rocks generally referred to as **zeugans** or, if the wind is predominantly one direction, **yardangs** (groovy ground) are seen. If the loss of material goes on over a sufficient area then deflation hollows can be formed. In terms of depositional features, the **sand dune** remains the classic desert symbol (although not the most common). Dunes are unstable masses of sand which will form wherever there's a sufficient drop in wind energy to deposit particles. This can be caused by an obstacle (obstacle dunes), bush or shrub (nebkha dunes), a depression (lunette), changes in wind direction creating linear dunes (seifs) or the classic crescentic-shaped barchans. However they're formed the process behind it is still the same.

Coasts are shaped almost entirely by water. Because they are at the boundary between land and ocean ecosystems they are places of high biodiversity: because of the erosive force of the oceans they are often dynamic places and because of the variety of rock types and climatic conditions etc. they are also some of the most varied places. The coast is not just a simple landform isolated from others. By being at the boundary of land and sea it is influenced by both. There is the energy from the ocean system

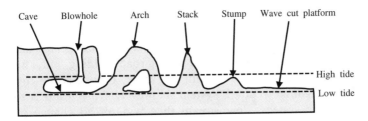

Figure 3.2.10 *Coastal features in cross-section*

($\Rightarrow$ oceanography) which meets the geological structure of the coast and a series of materials which might come from land or sea. To this can be added the moderating influences of coastal formations themselves. This can make classification difficult. Several classifications have been proposed based on sea-level variation, structural variation, plate tectonic activity and wave energy.

There are four main erosive processes involved at the coast. Wave action is where waves can force weak rocks apart (quarrying) or by scraping the surface (abrasion). Weathering, principally physical weathering, is also present: water layer weathering where wetting and drying or salt accumulation (or both) can break down rocks or subaerial weathering where normal processes like freeze-thaw action can occur. If the rocks are particularly soluble then solution processes can be seen (especially in limestone areas). Finally, bio-erosion is possible although how important it is remains a key question. It's probable that in higher latitudes it's relatively unimportant whereas by the tropics it becomes more evident. The most common landforms produced by erosion are cliffs and wave cut platforms (areas directly eroded by each tide's waves).

Although erosional landscapes are often seen, depositional ones may well appear to be the more numerous possibly because they represent a common view of the seaside. Landforms produced by deposition range from beaches and barriers (sand banks formed off the coast) to spits and tombolos (large, narrow areas of beach deposits where one end is joined to land). To illustrate the variety of landforms possible, Figure 3.2.10 brings together three aspects: the wave areas of the beach, erosional landforms and depositional landforms.

3.2.6 What is the extent of human-geomorphological interactions?

Finally, it is possible to assess the extent to which human action and geomorphological processes interact. Assume that the natural system is in some sort of systems balance with the rest of the environment: to this one adds some measure of human impact (Figure 3.2.11). Just like ecosystems, the physical environment will have some form of resilience and if the impact is small then nothing will happen. If the impact is too great then one of two things will happen. Human interference might cause disruption of the landscape (e.g. altering slopes during road building) but, after initial change (small landslips?) the system will settle. This could be considered part of the negative feedback system. However, if the result is to create changes in the systems downstream then widespread change might be seen. That would be positive feedback. What determines whether the impact is positive or negative? Although each site will be unique and therefore the response will be unique there are two factors that are important: the time-scale and spatial-scale of the impact. For example, digging a small ditch alongside a field would be an impact on the hydrological system but it would be so small-scale in both time (a few days' work?) and space (a few tens of metres?) that the effect would probably be negative (although similar small-scale changes in the Chalk Downlands of Southern England did lead to local flooding of a housing estate which shows how important the individual site factors can be). At the other end of the scale reversing river flows (as has been considered in Russia and China) could create global impacts that might be impossible to reverse.

What of the extent and amount of human impact on geomorphological systems? It's probable that there's no aspect of the landscape that does not have some form of human impact. To understand this there is need to consider a few examples. One of the most important aspects for urban areas is the supply of fresh water and yet getting hold of it can create serious problems elsewhere. Studies in Mexico City demonstrate that as population and water demand rose, so constant abstraction from the ground let to permanent subsidence. Whereas this may not be too severe, in other places such as Vienna, Italy, building on marshy areas has led to subsidence and flooding of historic buildings. Studies of erosion rates in Papua New Guinea have shown ten-fold increases from natural rates at the start of agriculture about 10,000 years ago which had risen to a 600-fold increase from natural rates in recent years which has significant results downriver. Rivers can be further affected by construction of dams. Even though the dams may hold vital water supplies their impact can range from subsidence due to pressure of water on the surface to reduction of up to 65% in downstream channel capacity.

The geomorphological impact on human activity divides into two: hazards and land-use management. Hazards are an obvious way in which geomorphology controls human action e.g. the volcanic eruption at Mount St. Helens in the USA or the effects of Hurricane Gilbert in the Caribbean. However, there's another side to this: the idea that geomorphological processes have to be taken into account every day. For example, the mudslides that have affected Vancouver, Canada for centuries have followed a particular set of paths. Rather than try to re-route the mudslides the authorities have banned building on them. In the UK at present, there is some debate over the way in which we should handle coastal erosion: to

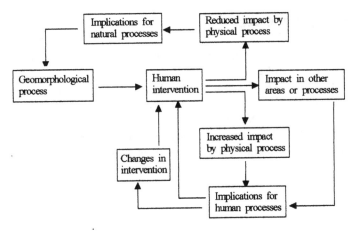

Figure 3.2.11 Impact of human intervention

spend vast sums on coastal defences or to relocate any people affected.

Most of the time, when people reacted to the physical environment in the past it was because they'd found a problem. It might have been difficulties with drainage or an unstable slope. Whatever it was the usual response was to act when the problem arose rather than to plan in advance. Today, the most common approach is to use Environmental Impact Assessment which tries to anticipate difficulties and suggest alternatives. In this way we can reduce our impact on the physical environment. As an increasing number of studies show us, the physical environment can be damaged as easily as the biological environment: we cannot afford to ignore it.

Questions

1. Using a map of your local area trace the boundaries of the main river basins. What common and contrasting features can you see?

2. Using maps 1:50,000 or 1:250,000 maps of contrasting areas trace out two catchment areas. Mark on all streams, lakes etc. Mark on contour lines at suitable intervals. Carry out the analyses mentioned in the text. How do the two compare? How can you account for any differences? What would happen in either case if human activity were to increase in the area? If you were in charge of the development for each area what would you recommend and why?

3. Why is there a gap between the erosion and the deposition curves on the Hjulstrom curve? On similar lines, why does the velocity need to increase for clay particles to start eroding?

3.2.8 References

Broadley, E. and Cunningham, R. (1991) *Core Themes in Geography: Physical*. Oliver and Boyd.

Collard, R. (1988) *The Physical Geography of Landscape*. Unwin Hyman.

Goudie, A. (1995) *The Changing Earth*. Blackwell.

Goudie, A. (1993) *The Nature of the Environment*, 3rd edition. Blackwell.

Hansom, J.D. (1988) *Coasts*. Cambridge University Press.

Waugh, D. (1995) *Geography*, 2nd edition. Nelson.

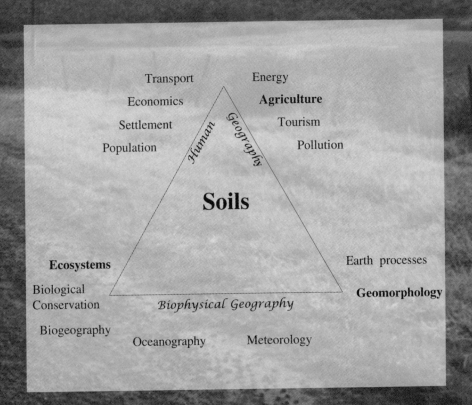

Transport   Energy

Economics   **Agriculture**

Settlement   Tourism

Population   Pollution

Human Geography

Soils

Ecosystems   Earth processes

Biological
Conservation   **Geomorphology**

Biophysical Geography

Biogeography

Oceanography   Meteorology

Subject overview: soils

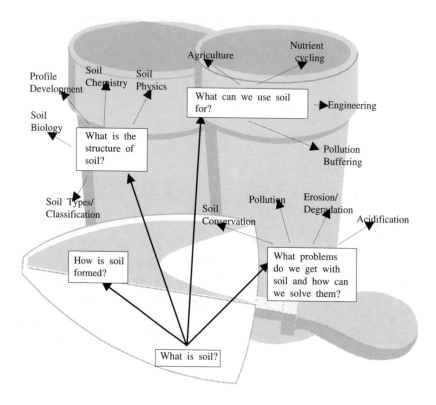

1. Soil – a surface layer (typically 1 m deep) composed of mineral particles, and organic matter.
2. Soil is an ecosystem in its own right.
3. Soil is formed by the action of weathering, climate, relief and organisms on the parent material over a period of time.
4. Soil texture is micro-scale and controls nutrient cycling porosity; structure is meso-scale and controls drainage.
5. Soil leaching and eluviation are major factors in nutrient distribution.
6. Soils are described using a standardised profile. British soils can be divided into four main groups – brown earths, podsols, gleys and rendzinas.
7. Major soil uses include agriculture, nutrient, cycling, engineering and pollution buffering.
8. Soil problems occur when activity exceeds the soil's ability to restore itself.
9. Many problems with soil usage could be solved using techniques already available.

3.3.1 Introduction

Life on Earth depends upon plants producing oxygen and recycling carbon dioxide. Without this, we wouldn't exist. For plants to exist they need to grow in a suitable place (soil) but rarely are soils deeper than 1 m. The radius of the Earth is about 3000 km. Soil is therefore only 0.000033% of the radius. A minute fraction and yet it supports life on the planet. So we need to look in much more detail about this thin layer, see how it works and what we do to it.

3.3.2 What is soil?

It is important to distinguish between a proper soil and a layer of loose particles. There are several processes that take a loose sediment and turn it into a soil. This is similar to succession in ecology: soil is the climax product. Soil can be defined as: *a surface layer composed of mineral particles and organic matter*. However, that's not the complete story: soil is not just an collection of bits it is an ecosystem in its own right.

This special ecosystem also provides shelter and food to all the other terrestrial ecosystems. Soil is the link between all the other ecosystems. The three key functions of soil are: to provide nutrients and support/shelter for other ecosystems, to act as a major supplier of decomposers vital for the recycling of nutrients and to act as a 'buffer' between several other systems (a buffer is something which reduces the impact of an input). For example, it can slow down the flow of water in the hydrologic cycle, reduce the rate of temperature changes and even reduce changes in acidity.

> **Tip**
>
> One common mistake is to say that plants (such as flowers and trees) are part of the soil ecosystem. They're not! There is a special set of plants and animals that live in the soil (rather than using it for food and support).

3.3.3 How is soil formed?

Although soil is being formed all the time the easiest way to think about its formation is to start with bare rock (usually called *parent material* by soil scientists). Soil development (also called **pedogenesis**) can be seen as being very similar to plant succession. Consider Figure 3.3.1. The three drawings show a section of rock as its surface becomes soil. In the first case, there is no soil: only a few mosses and lichens can colonise the surface. They start to break down the rock and, along with their organic remains, start to build up a small layer of 'soil' (it's not true soil yet). The next picture shows the plant succession to have moved to grasses and shrubs. This continues to break up the rock and add more organic material - the soil ecosystem is developing. The final picture shows the fully developed soil which can support a woodland. Of course, in reality the system is not that clear but the basic principle is there: the gradual development of the soil ecosystem through continued breakdown of rock, the addition of organic material and the introduction of soil animals and

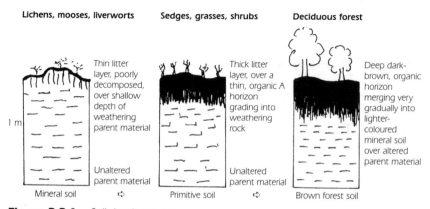

Figure 3.3.1 Soil development

plants. This process can take a very long time to achieve. British soils are considered to be very young at a mere 10,000 years; some tropical soils may be over 1-2 million years old! Because of the factors which influence soil development, one can get an enormous variety even within a very small area.

There are a number of factors involved in soil formation. Firstly, there is **weathering** – the breakdown of parent material to a series of mineral particles. Of the three types of weathering only physical and chemical weathering have any real influence. What is there depends on the rock type involved. Essentially, weathering can have three effects: it breaks up rocks to allow plants to grow and add organic matter to the soil; it can lead to changes in the types of minerals as they are altered during weathering or it can allow some chemicals to be lost. Of course what type of weathering is present and what happens to the rock depends on the **climate**. This can be a major influence e.g. lack of water in a desert restricts the soil's development. Some soil researchers argue that the effect of climate is so great that one can divide the world's soils into three: *zonal* (influenced by climate); *intrazonal* (influenced by parent material) and *azonal* (restricted soil development). Thirdly, the **relief** or topography of the ground can have a large effect. For example, on a steep slope much of the soil can be washed away to the bottom giving poor soils on the hillside and deep, rich (even water-logged) soils in the valley. The best example of this is called a **catena**. Take a chalk downland slope (Figure 3.3.2). On the top, the soil is easily eroded leaving a thin, often poorly developed soil. The same process affects the slope. When the valley floor is reached there is a rich, deep soil where all the up-slope eroded material has been deposited. Since slopes are common features it follows that catenary sequences are commonly seen, indicating the importance of relief to soil development. **Organisms,** are

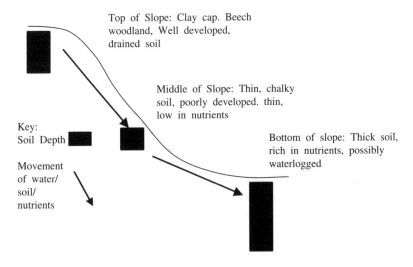

Top of Slope: Clay cap. Beech woodland, Well developed, drained soil

Middle of Slope: Thin, chalky soil, poorly developed. thin, low in nutrients

Key:
Soil Depth

Bottom of slope: Thick soil, rich in nutrients, possibly waterlogged

Movement of water/ soil/ nutrients

Figure 3.3.2 Development of a catena

the fourth factor, not just in the soil aiding the breakdown of organic material but also acting to break down parent material (e.g. rabbits). **Time** is often the most neglected influence and it can be the most important. The longer the soil has to develop the more mature its characteristics. Thus British soils with their brief timespan of 10,000 years may be no more than 1 m deep whereas tropical soils with hundreds of thousands of years can be over 30 m deep (some are so well developed that they are mined for their mineral content e.g. bauxite).

> **Tip**
>
> Soil is constantly being developed and eroded: both activities occur together. Therefore soil depth = rate of formation – rate of erosion.

These factors are important because they can be seen in more than one situation. Here they are linked to soils but Hans Jenny also linked them to ecosystem development. The **Jenny equation** D = CLORPT (where CL = climate, O = organisms, R = relief and water, P = parent material and T = time.

> **Tip**
>
> Remember CLORPT. Use it as a basis to compare different soils.

3.3.4   What is the structure of soil?

All the factors noted above can act in varying amounts in different areas to give a great variety of soils. Despite this variety there is a basic structure to soils the world over. In other words we can look at soils from North America or Europe and still find aspects in common. This common pattern is called the **structure** of the soil. Structure, the linking of particles into larger pieces, often called **peds**, is a meso-scale feature which distinguishes it from **texture** (a micro-scale feature relating the proportions of different-sized mineral particles). Four main types of peds have been described: granular, platy, blocky and prismatic.

It is possible to divide soil structure into three: *physics* (covering the arrangement of soil particles); *chemistry* (the reactions between soils and a range of inputs) and *biology* (the soil ecosystem). These three aspects combine in different circumstances to give a **soil profile** (a cross-section through the soil). Finally, these profiles can be grouped together to create a soil classification.

The physics of soil refers to the particles that make up soil and the way they are arranged. For example, to build a road one must know what the soil is and how it will react to the increase of weight on it – get it wrong and the road could collapse! The physical part of soil is made up of four components in varying proportions: mineral matter (40–60%), water (20–50%), air (10–25%) and organic matter. Soil scientists are usually concerned with three major mineral components: sand, silt and clay. The proportion of these can be used to classify soils using the **tri-plot diagram** (see Figure 3.3.3). Still at the level of the individual particle it is important to realise that soil is not just minerals packed together. Each particle is an uneven shape and so the packing of these particles is bound to give rise to spaces or pores. These pores can be filled with water and air and are important for both the soil's organisms and those from other ecosystems using the soil. The amount of pore space gives an indication of the way in which water can pass through soil. For example, large pore spaces can lead to a soil being well drained (important for farmers) but if it is too large it can be excessively drained and lead to a poor soil (such as is found on gravel). Very small pore spaces (e.g. clay soils) promote water-logging. The air in the soil is not like that in the atmosphere but is just as important. Soil air is lower in oxygen and higher in carbon dioxide but the exchange of soil and atmospheric air at the surface is necessary for the organisms to function.

Two more concepts cover the key physical aspects of soil: temperature and water flow. Soil temperature is important for soil organisms, seedling and root growth. Typically, soil temperature will differ from that of the surrounding air because of solar radiation, heat loss by soil, water evaporating and the thermal properties of the soil. Also important is the reflectivity of the soil (called its **albedo**) which can be as high as 95% on fresh

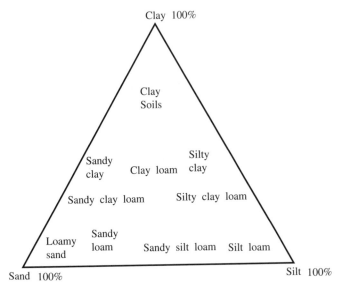

Figure 3.3.3 Soil tri-plot

snow and as low as 5% in a conifer wood. As the flow of water can control the amount of plant growth in the soil it is regarded as a key component ($\Rightarrow$ hydrologic cycle). Four elements of the hydrologic cycle affect soils most: infiltration (the rate at which water flows into the soil), leaching (the removal of soluble plant nutrients through the soil caused by excess rain in well-drained soils), eluviation (similar to leaching but deals with insoluble particles) and illuviation (the deposition of particles). In addition, there's a term restricted to soils that is vital in understanding irrigation: **field capacity**. This is the amount of water a soil can 'hold' under natural conditions. Imagine being in a bath and holding a sponge under water. Squeeze until all the air is expelled. Soil under such a situation would be called waterlogged (and in poor condition – with no air, most organisms can't live). Take the sponge out of the water and let the excess drain away (throughflow in hydrologic terms). What remains is field capacity. If the sponge is squeezed out it is possible to reach a stage when no more can be squeezed out. This is a soil at its wilting point – plants can get no more water and start to wilt. Irrigation aims to keep a soil just below its field capacity.

Soil chemistry deals with reactions. One of the most important concepts in soil chemistry is the way in which nutrients are held (or not) in the soil. When parent material decays it releases large numbers of charged particles (called ions: cations – positive and anions – negative). These ions can bond with clay particles and with organic material (humus). This **clay-humus micelle** is a key component in determining soil fertility. Cations bond with the micelle leaving the anions to be leached out. Two vital cations are the major nutrients phosphorous (P+) and potassium (K+). Other cations (usually in far greater amounts but not as vital to plants) are calcium, magnesium and sodium. Note that nitrate fertilizer (the third plant nutrient) is an anion which will not bond: it is washed through the soil often leading to water pollution (eutrophication). Two points are worth noting here. Firstly, the amount of cations a soil contains is a measure of its **cation exchange capacity** which defines its ability to store nutrients. Secondly, there's a link between the exchange capacity and acidity. Under conditions of high acidity (pH <5) a range of cations can be released including aluminium which is a major component in acidification of lakes and streams and a major fish toxin.

Another important concept is acidity (pH). Many reactions in the soil are controlled by acidity including the release of nutrients. Below pH 5 many nutrients are no longer available for plants to use which is why acid soils tend to be less fertile. Highly alkaline soils will also have problems. The ideal range is probably pH 6–7.

Even some of the poorest soil can be teeming with life – with tens of millions of individuals per hectare. Most of these organisms are decomposers. It is their niche that provides the energy for the soil ecosystem because, of course, sunlight is missing. With their diversity, soil organisms play an important role in the environment and in crop production. Not all the organic material in soil is living. Decaying/decayed organic matter is broken down by the decomposers and is incorporated into the soil as humus. Some humus e.g. that found in deciduous forests can support an abundance of soil organisms. Such **mull humus** as it is called is in contrast to the **mor humus** of coniferous areas which is hard to decompose and contains less usable material.

It is now possible to put the three elements of physics, chemistry and biology together to describe the various major soil types. The basic unit for soil description and classification is the **profile**. This is a section taken to show all the major divisions from plant life growing in the soil to the parent material. An ideal profile can be described (see Figure 3.3.4 and below): it is then possible to link all soils to that profile. The advantage of doing it this way is that it will provide a standard reference against which others can be measured. The first thing to note is that the profile is divided into a series of soil **horizons** (the term layers is reserved to describe the structure of decomposing organic matter on the surface e.g. leaf litter). Although the precise notation of soil profiles might vary there is general agreement about the following descriptions:

1 **O layer** – organic material, usually dead organic material and waste. This is the organic nutrient supply for the soil. This material is gradually broken down by the soil decomposers (such as fungi and earthworms) and then incorporated into the soil horizons. It can be divided

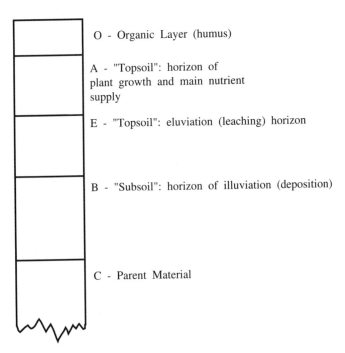

O - Organic Layer (humus)

A - "Topsoil": horizon of plant growth and main nutrient supply

E - "Topsoil": eluviation (leaching) horizon

B - "Subsoil": horizon of illuviation (deposition)

C - Parent Material

Figure 3.3.4 Ideal soil profile

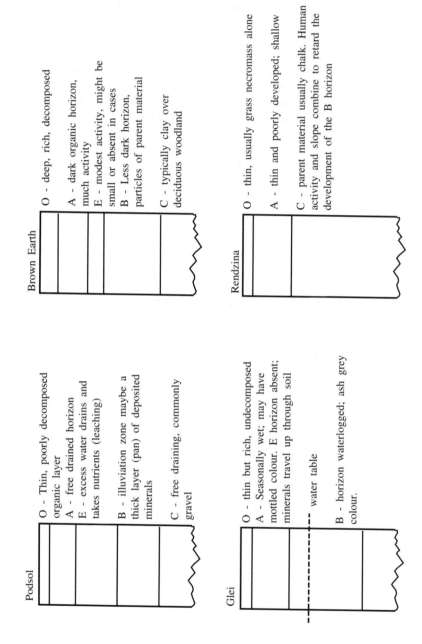

Podsol

O - Thin, poorly decomposed organic layer

A - free drained horizon

E - excess water drains and takes nutrients (leaching)

B - illuviation zone maybe a thick layer (pan) of deposited minerals

C - free draining, commonly gravel

Glei

O - thin but rich, undecomposed

A - Seasonally wet; may have mottled colour. E horizon absent; minerals travel up through soil

-- water table

B - horizon waterlogged; ash grey colour.

Brown Earth

O - deep, rich, decomposed

A - dark organic horizon, much activity

E - modest activity, might be small or absent in cases

B - Less dark horizon, particles of parent material

C - typically clay over deciduous woodland

Rendzina

O - thin, usually grass necromass alone

A - thin and poorly developed; shallow

C - parent material usually chalk. Human activity and slope combine to retard the development of the B horizon

Figure 3.3.5 Major UK soil types

into two – the fresh material which has not been broken down (called the O1 layer) and the underneath O2 layer where decomposition has started. It is not humus until it has been taken into the soil.

2 **A horizon** – the first true soil layer. This is the horizon of maximum humus uptake – any nutrients in the O layers will be washed or carried into this area. It is also the horizon of maximum plant and earthworm activity. Often, gardeners refer to this layer as the 'topsoil'.

3 **E horizon** – which stands for eluviation. In this horizon there is the maximum amount of organic material taken (usually leached) down the profile to the lower horizons. If the A horizon is where the humus is incorporated into the soil, the E horizon is where it is transported. In some soils this can be so extreme that the E horizon is left with virtually no nutrients.

4 **B horizon** – horizon of illuviation, or deposition of inorganic and organic chemicals. Of all the horizons this is likely to be the most varied. Consider it to be the meeting place of chemicals washed down and decomposing parent material – a dumping ground which can control the whole soil structure. In some areas this horizon can be so hard from chemical deposition that it can form a pan: in other places it may be so thick as to be worth mining (e.g. tropical soils with bauxite – aluminium ore).

5 **C horizon** – the parent material, which may be a rock or sediment such as glacial material. The importance of this layer is often underestimated. Whereas the plants can exert a great force (especially in forests) and climate can be a controlling factor (as in tropical soils) the basic chemical make-up of the parent material determines what is available in the first place.

> **Tip**
>
> Not all soils have all horizons or have developed them in the same way. Go back to look at the soil-forming factors: each one can exert a different influence.

This gives us a framework which we can use to describe soils. In the UK and in similar areas one can find four major soil types. Look at Figure 3.3.5 for a diagram of the profiles:

1 **Brown earth**. The most common forest soil. Note the humus-rich A horizon which has accumulated the organic material from decomposing leaves. It would tend to be the deepest of the soils: tree cover would reduce the amount washed away. It would also tend to be the most fertile but this fertility can easily be lost as happened when the Neolithic people deforested the Chalk Downlands of Southern England.

2 **Podsol**. A soil which is dominated by eluviation. Found on excessively

free-draining sites such as gravels the pan is a characteristic feature. Note the A horizon which is almost cleared of nutrients by leaching (the washing of nutrients down the profile). The characteristic orange colour of the B horizon shows where most of the iron and aluminium cations have been deposited. The presence of these two chemicals (both toxic in large amounts) means that the soil will only be found in the most acidic of areas and that it will have a very low fertility.

3 **Gley** (or glei) is a common example of a waterlogged soil. The B horizon is almost always below the water table giving rise to reducing rather than oxidising conditions and to a mottled appearance to the profile. This soil would be common near rivers and similar boggy areas. It should not be mistaken with the peat soil which is a special example of an almost entirely organic soil.

4 **Rendzina**. Less common on a world-wide scale than the other three this represents an altered soil which has not reached maturity. In the UK it is virtually restricted to chalk areas. The rendzina shows what happens to soils which are altered highly by human activity. Once the trees had been cleared on the downland, the weather readily eroded the soils on the hilltops and slopes. This has lead to an 'immature' soil (i.e. not fully developed) where only the A and C horizons can be readily detected. It is common to find chalkland areas developing cantenas.

> **Tip**
>
> Although the examples above are from British soils there are numerous other classifications in common use. American, Polish and Russian scientists have all been active in this field; each with their own system.

3.3.5 What can we use soil for?

The simple classification described above was an attempt to show how similar soils could be grouped together. Although it is a very useful tool it is not the central question of millions who rely upon soils and their properties around the world. It's a simpler question – what can we use soil for? This section takes a brief look at some of the ways in which we can use soil. There are four areas to explore: agriculture, nutrient cycling, engineering and pollution buffering.

Agriculture has the aim of producing a higher biomass from an area than could be found naturally. Soil is the medium through which this is done. In other words, soils are managed to create the best conditions for the plants to be grown. This involves a number of practices, each one aimed at improving what already exists: cultivation, fertilizer use and irrigation are the main ones. Each of these techniques mirrors a natural process to provide what nature didn't.

Cultivation covers a series of methods which tries to improve the physical conditions of the soil. By breaking up the surface (ploughing) or adding chemicals (liming) to increase the pore space (and therefore aid water and air flow through the soil) the farmer is trying to create a crumb structure – a particular texture which balances water and air flow to maximise plant growth. By breaking up the peds to improve drainage and aeration one is dealing with improvements to structure. Cultivation also aims to turn the soil and so bury weeds and their seeds. There is some argument about this because although it is easy to understand – bury weeds and they'll die – the effects are not at all clear cut. Many weeds will only grow after the soil has been disturbed and some seeds will germinate only after they've been disturbed. Some farmers are trying zero ploughing or chemical ploughing (i.e. no disturbance or using sprays to kill weeds) to get around these problems and to ensure the soil is not compacted. In addition, there is the need to create particular patterns in the soil which might be needed by crops or help reduce problems of cultivation. An example of the first might be the creation of ridges to allow potatoes to grow. An example of the second would be the use of contour ploughing (ploughing parallel to the contours) or terracing to slow down soil erosion on slopes.

Fertilizers are an example of control of the chemical environment. Too little and the crop will hardly grow at all. Once more fertilizer is added then plant yield will rise. At a certain limit, a great deal of fertilizer will result in only a small increase in output. This is the optimum: any more and the yield will decline because of the excess of chemical in the soil (law of diminishing returns).

Finally, irrigation: the practice of reducing the soil's deficit for water. Today, an increasing number of areas see that their needs for water are outstripping the natural supply. Since 1950 the area of irrigated land in the world has virtually tripled so that in 1985 some 271 Mha were involved (that's almost 3 million square kilometres – one-third the size of the USA or 12 times the UK). Irrigation aims to counter the poor drainage effects of the natural soil. It has been estimated that even in places like the UK with a relatively good rainfall pattern up to 30% of the land could benefit from irrigation.

The second example is **nutrient cycling**. One of the uses of soil is to act as a means of transferring nutrients (plant foods). Consider Figure 3.3.6. Here, the nutrient cycle is linked into other aspects of the soil and associated ecosystems. The soil provides a store for nutrients from plant decay (through humus) and a source of new chemicals (from its own parent material). Plant nutrients are usually divided into two groups: **macronutrients** (or essential nutrients) – those required in large amounts for plant growth and **micronutrients** (or trace elements) – those needed in small amounts to aid specific aspects of plant development. In the former group, nitrates, phosphates and potash are the 'big three' whilst in the

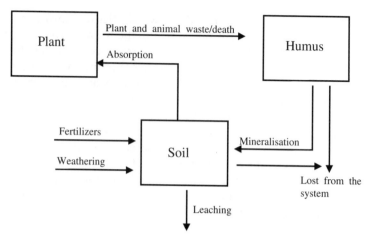

Figure 3.3.6   *Soil nutrient movement*

latter group a wide range of requirements, led by copper, zinc and calcium, are needed (⇒ ecosystems).

Engineering is probably not the first use of soil that one could think of but it is an important element in many countries who rely upon soil properties to build structures. In countries where water supply is irregular or low, then soil can be used to construct a series of water conserving measures and reduce soil erosion. Several properties of soil are crucial if this is to be successful. If the soil can retain water, or reduce the flow through the system then it could be good enough to build dams. In terms of particle size – a high percentage of clay would indicate a soil with low permeability which would be good for construction purposes. The compaction characteristics are important if the soil is going to be used in dam building and road construction.

Pollution buffering might seem like the odd one in the series but the value of soil in this capacity is becoming recognised more and more. Water flow through the soil is neither total nor instantaneous. This means that soil can act as a medium that slows down the release of a pollutant. In doing so it might even allow a complete change to take place. One example of this is in sewage treatment (⇒ pollution). In some rural sewage works the final stage of treatment can be via a grass bed. This is a grassy hollow into which the virtually clean water flows. The slow filtration of water through the soil and parent material not only slows down the process but even ensures that any remaining pollutants decompose and are therefore safe.

3.3.6 What problems do we get and how can we solve them?

This section looks at what happens when things go wrong and what we can do so stop it. There are several examples that we could use to illustrate this aspect of soils: pollution from acid rain and heavy metal waste, declining fertility through overuse, problems of compacted soils or degradation from overuse in marginal areas but perhaps the best case is that of soil erosion.

Why soil erosion? Because it covers a wide range of issues. It is widespread – found in virtually every country. It can have a serious impact on agriculture. It can highlight problems facing the developing world whilst still be able to be related to developed world issues. One obvious point is to define soil erosion. Two conditions need to be present: removal must exceed formation and there must be some human agency involved. Thus we can get natural soil erosion but it rarely becomes a problem unless it affects human activity or if human activity 'helps' the process. Figure 3.3.7 shows those areas of the world most seriously affected by soil erosion.

Soil erosion is caused by two agencies – water and air. Water erosion is caused by two effects: the impact of raindrops and the run-off of water down a slope. Raindrops possess a certain amount of kinetic energy. When they strike a surface this energy is dissipated. The net effect is to loosen soil particles. Water that cannot penetrate the soil will then run-off either as a sheet over a large surface or, more usually, via the creation of rills and runnels (very small 'valleys' on the slope – maybe only a few centimetres across). These small features should not be underestimated. In some

Figure 3.3.7 Areas of the world most affected by soil erosion

Southern African states, for example, a small channel can turn into a 1 or 2 metre wide feature virtually overnight. Wind erosion affects dry soil. If the velocity is high enough and/or the particle small enough then there is sufficient energy for it to be transported. Wind erosion is greatest in areas with little vegetation cover and with high wind speed.

How can people assist these natural forces? Generally by misuse of the land but that can make it seem too simple. For example many of the areas with greatest erosion are also those with the poorest communities least able to try to stem the problem. Some writers argue that soil erosion is a political issue because of the forces that seem to constrain human action. The main causes are the destruction of woodland which exposes a soil previously sheltered to the full effects of climate; the cultivation of grasslands, marginal areas and steep slopes where the already sparse vegetation is removed and grazing particularly in marginal areas. Some forms of nomadic pastoralism (e.g. goat herding) have been seen as a major factor in the increase in desert in the Sahel region.

Erosion reduction techniques can be grouped into three classes. Firstly, there are biological methods. With the aim of keeping plant cover at a maximum when erosion is at its highest then one can use crop rotation, leave vulnerable areas permanently covered or plant/harvest in strips rather than areas. Hedgerows are also very useful in reducing erosion (in addition to their wildlife value). Secondly, there are cultivation methods – using ordinary farm equipment in new ways. For example, rather than ploughing up and down slope, contour ploughing (going parallel to the slope) reduces the velocity of water down the slope. Another method would be to use minimum tillage (mentioned above in connection with soil compaction) which would reduce the amount of fresh soil exposed to erosion. Finally, there are mechanical methods, using techniques which regrade the land of which terracing is perhaps the best known. Another method which has only recently started to gain some acceptance is permaculture (permanent agriculture). Here the aim is to reduce impact on the area whilst maximising output. To do this many soil conservation techniques are used such as covering the soil or planting parallel to contours, building small terraces to catch the water for one tree for example.

Questions

1. What would happen to other ecosystems if soil didn't perform buffer functions?
2. What are the implications of soil variations in small areas for farmers? How could it affect farming activities?
3. Some argue that it is better to give fertilizer to a nutrient-poor field than a nutrient-rich one. What arguments could be put forward to support this case?

3.3.7 References

Charman, P.E.V. and Murphy, B.W. (eds) (1991) *Soils: Their Properties and Management*. Sydney University Press.

Cresser, M., Killham, K. and Edwards, T. (1993) *Soil Chemistry and its Applications*. Cambridge University Press.

Killham, K. (1994) *Soil Ecology*. Cambridge University Press.

Wild, A. (1993) *Soils and the Environment*. Cambridge University Press.

White, R.E. (1987) *Introduction to the Principles and Practice of Soil Science*. Blackwell Scientific.

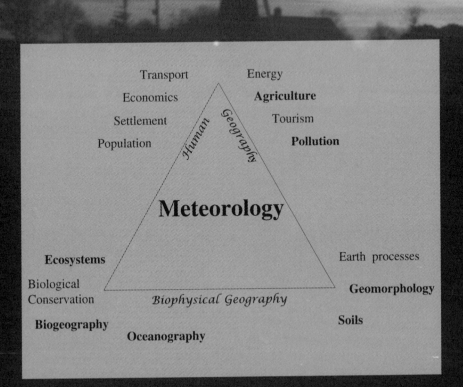

Transport Energy
Economics **Agriculture**
Settlement Tourism
Population **Pollution**

Human Geography

Meteorology

Ecosystems Earth processes

Biological
Conservation **Geomorphology**

Biogeography Soils

Oceanography

Biophysical Geography

Subject overview: atmosphere

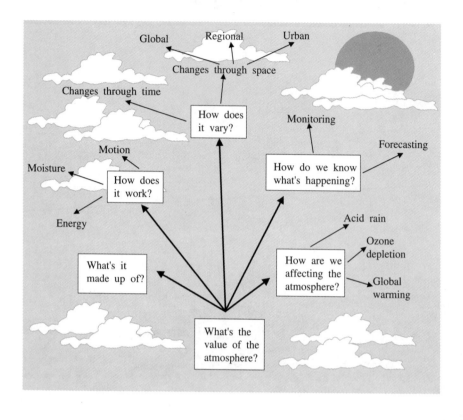

1. The atmosphere is of central importance. It transports energy and supports life.
2. The four key gases – nitrogen, oxygen, argon and carbon dioxide make up 99.98% of the atmosphere. The remaining substances, including pollution, often cause the major changes we see.
3. Atmospheric movement is based on differential heating – the forces caused when one area becomes hotter than another.
4. One of the key principles in understanding weather phenomena is the relationship between water vapour and temperature.
5. To understand global weather/climate types we need to construct a model of general atmospheric circulation. The way in which these global patterns are made helps us predict future climate trends and helps us understand air pollution.
6. 'Microclimates' in our larger cities have a profound effect on how we live by altering patterns of rainfall, snow, fog etc.
7. Today, we are affecting the climate in numerous ways. Some of these might be limited and local but some are potentially worldwide e.g. global warming. Usually our knowledge is not sufficient to predict what could happen.

3.4.1 Introduction

Our use, or more accurately misuse, of the atmosphere is of keen interest because it affects all of us. Global warming, acid rain and ozone depletion have become topics of concern throughout the world. This chapter outlines the basic concepts upon which our understanding of the atmosphere is based.

3.4.2 What's it made up of?

The atmosphere is a mixture of a range of substances (see Table 3.4.1). The four key ones are gases: nitrogen, oxygen, argon and carbon dioxide which together account for about 99.98% of the atmosphere and which appear to be present in roughly the same amounts throughout the atmosphere. Other gases exist in minute trace amounts. In addition to these gases there is water vapour. There is also a range of solid particles which can be from natural (sea spray, volcanoes etc.) as well as from human sources (burning and industrial processes).

Table 3.4.1   Composition of the atmosphere

Component	Percentage
Nitrogen	78
Oxygen	21
Argon	1
Carbon Dioxide	0.04
Ozone	0.00006

Tip

Many of gases and solids that can change the atmosphere are present in minute quantities. It doesn't take much to trigger a change.

This mixture of gases, liquids and solids shows little variation in composition either horizontally or vertically. However, there are some differences which are important for atmospheric processes. One of the more important is the influence of height. Water vapour is typically concentrated near the surface whilst ozone (produced naturally by the breakup of oxygen molecules by solar ultraviolet radiation) is found between 15 and 35 km from the surface. Latitude and season are also causes of variation. For example, ozone tends to concentrate around the poles in spring; water vapour is related to temperature and so you'd expect this to change with season. Finally, there's time. The greatest changes have been in human-made sources of pollution such as carbon dioxide although studies have shown similar variations to have been present from natural sources in the last Ice Age.

Studies have shown that there are a series of atmospheric layers each surrounded by a thin area in which rapid change occurs (Figure 3.4.1).

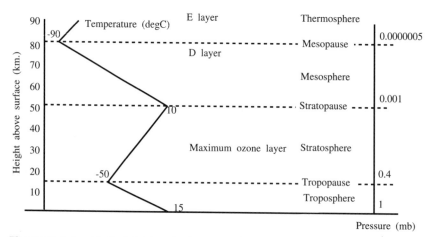

Figure 3.4.1   Vertical divisions in the atmosphere

The troposphere, nearest the Earth, contains the majority of gases and meteorological phenomena (like clouds, rain etc.). At about 8–17 km above the surface there is the tropopause forming a cap to the troposphere. Above that is the stratosphere. In this layer the key feature is the concentration of ozone at about 25 km. Again, there is a 'cap' at about 50 km – the stratopause. Beyond this you are getting into the realms of space and definitions of layers depend on which system you use. Most accept that the mesosphere is next followed by the thermosphere which, by now, is very low pressure and made up of atomic oxygen rather than the molecules of lower layers.

3.4.3   How does it work?

Energy distribution and transport are the most fundamental parts of the atmosphere. Without energy transport the equator would be, almost literally, boiling whilst the poles would be surrounded by carbon dioxide ice rather than water ice (and probably cold enough to give pools of liquid oxygen!). That this doesn't happen is down to the effects of the moving atmosphere (and ocean currents) (⇒ oceanography) which transports energy around the world. We experience this energy as wind and rain etc. It's been estimated that this movement of solar energy accounts for 99.7% of energy transfer in the atmosphere.

The Earth is being heated by the Sun but it isn't getting any hotter (or colder). This is due, in part, to the **energy budget** (see Figure 3.4.2). Solar energy (or radiation) reaches the upper atmosphere. This radiation has a range of wavelengths (in the electromagnetic spectrum) and differing amounts of energy. High energy ultra-violet reacts with the ozone layer. Some energy is reflected off clouds and other surfaces (variations in which

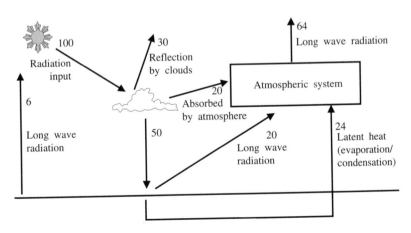

Figure 3.4.2   Energy budget

The energy in the atmosphere can be seen in four forms: kinetic (the energy gained from motion); potential (energy that could be gained by a move of position); latent (energy gained or released during the processes of condensing and evaporating, respectively); and thermal (stored in an air mass because it's warmer than its surroundings).

we refer to as **albedo** – the reflectivity of a surface in percentage terms). The atmosphere gains energy and will start to move around. Some energy reaches the surface (**insolation**) and heats up the surface. If this gets hotter than the surrounding atmosphere it will give out heat energy. This will not be the short-wave energy of incoming radiation but a longer wave energy (see Figure 3.4.3). As with other aspects of the atmosphere, there are variations. Firstly, there's solar radiation. The sun alters its output and the Earth its orbit. Variations also occur with altitude and the length of the day. Next there's atmospheric variations – some gases have a greater effect than others. Clouds and dust can give a significant variation. Thirdly there is the effect of latitude – more radiation reaches the equator than the poles. Finally, there is the distribution of the land masses. The direction the land is facing (called its aspect) also determines the radiation received in a given area. In the northern hemisphere, north-facing slopes gain less than south-facing ones.

Motion is vital to the atmosphere for without it solar radiation could not be distributed. There is one basic principle: motion will only occur

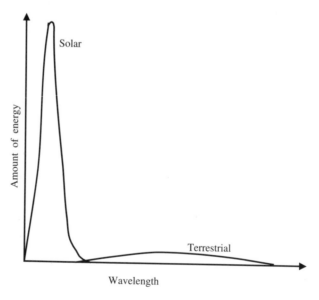

Figure 3.4.3   Solar and terrestrial radiation

when there is a pressure difference between two places. Since we're dealing with a three-dimensional layer it follows we can look at two elements of motion: horizontal (by far the greater force) and vertical. Horizontally, there are four forces which cause the motion of air to vary. **Pressure-gradient force** – the greater the difference in pressure between two places, the greater the pressure-gradient force. To understand this force, think about contours on maps. If there are few or no contours the land is flat (and there's little movement). If the contours are numerous the place is hilly and there can be a great deal of movement. The **Coriolis force** relates to the movement of air masses. To understand this one has to look at the law of conservation of angular momentum: keep the mass the same and as the radius varies the velocity changes. Imagine an air mass at the equator. It has one velocity. Further north (or south) the radius will be less and the velocity greater. But, the Earth will be rotating at the same speed and so an observer will see the air mass rotating to the right in the northern hemisphere. Thirdly, the **centripetal force**. Any spinning object will try to move away from the centre (the centrifugal force). This is kept in balance by the force pulling it inwards – the centripetal force. Generally this has little impact except in extreme circumstances such as tornadoes. Finally there are the **frictional forces** – very important near the surface because they act to slow down movement. This force can be great enough to overcome other forces and alter the direction of the movement (i.e. surface direction will differ from direction at height).

It was once thought that vertical forces were of little importance but since new models of the atmosphere have been made it's possible to appreciate their contribution. Three forces help explain the variations recorded. In terms of **convergence and divergence**, air may move towards or away from a place. This would give rise to more air than one would expect (a convergence of air) or less (a divergence) based on the pressure at that point. These phenomena are seen as being important in our understanding of the formation of depressions and high pressure areas (see below). Another force is **vertical motion**, the movement of air resulting from a vertical pressure gradient. Thirdly, **vorticity**, the 'spin' on an air mass, is related to the Coriolis force. This can influence the amount and direction of air movement over such places as mountain ranges. All of these forces can be used to model the atmosphere. Since they use well-known basic equations it is possible to set up a computer to produce detailed maps of the atmosphere. The General Circulation Model or GCM is one of many models that have been constructed and refined in recent years. All the forces that have been described above have come from basic physics, the equations for which have been known for many years. This means that the atmosphere can be reduced to a series of mathematical statements. Put these in a model and run it and you will get an approximation to the real atmosphere.

The final part of this section involves moisture. Although present in small quantities it can have a dramatic effect. Water is present in the atmos-

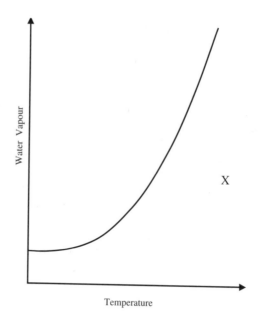

Figure 3.4.4 *Variation of relative humidity and temperature*

phere as a gas–water vapour. The amount of vapour that air can hold varies with temperature (Figure 3.4.4). Pick a point that's below the line of the graph (say, X). At this temperature there's less moisture than the air can hold (i.e. **unsaturated air**). Cool the air mass but keep the water vapour the same value (go horizontally to the left). You'll reach a point where you touch the graph. Now the air is saturated (and the point you've reached is called the **dew point**). Cool the air mass further and the vapour will be more than the air can hold and it will condense to form clouds.

> **Tip**
>
> Remember the water vapour/temperature relationship – it's one of the key principles.

Imagine this air mass is isolated from its surroundings. Something causes this air mass to rise and in doing so it can expand and cool. A graph of this air mass' temperature against height would be a straight line like Figure 3.4.5a. This is the **dry adiabatic lapse rate** or DALR (dry, because it's not saturated, adiabatic because it's an isolated air mass and lapse rate to describe the gradient of the line). With a saturated air mass Figure 3.4.5b will be produced. The **saturated adiabatic lapse rate** (or SALR) graph is curved because as the air is cooled it becomes saturated, water vapour is turned to water with a gain in temperature (from the latent heat of condensation). A typical cloud formation graph (Figure 3.4.5c) is

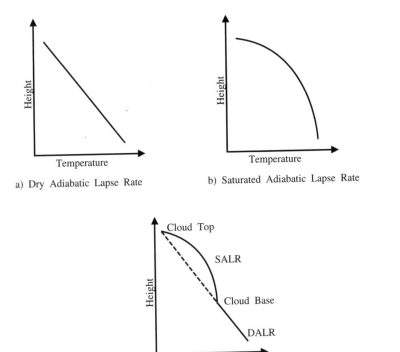

a) Dry Adiabatic Lapse Rate

b) Saturated Adiabatic Lapse Rate

c) Combined Adiabatic Lapse Rates

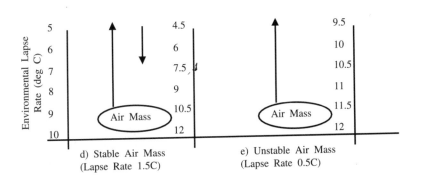

d) Stable Air Mass
(Lapse Rate 1.5C)

e) Unstable Air Mass
(Lapse Rate 0.5C)

Figure 3.4.5 Lapse rates

produced by the combination of the two graphs. The air mass cools by the DALR until it reaches its saturation point. Condensation occurs and the line changes to the SALR. Eventually a situation is reached where the air mass has no more water vapour to lose or where the surrounding temperature equals that of the air mass. This shows the top of the cloud.

There are two more conditions. Add a surrounding temperature change with height (the **environmental lapse rate** or ELR). In Figure 3.4.5d the air mass cools faster than the surroundings and so it sinks back to Earth. This is referred to as a **stable air mass**. Alternatively, Figure 3.4.5e shows what happens when the SALR is less than the ELR – the mass keeps rising which gives an **unstable air mass** (and probably a thunderstorm!). It's not just a question of cooling an air mass to get condensation: we need something for the condensation to form around – condensation nuclei can start the process.

Water droplets will eventually become rain but the way in which this occurs is the subject of two competing theories. In the **collision theory**, water droplets collide and thus join together. Once big enough they will be able to fall as rain. In the **Bergeron–Findeisen theory**, ice crystals are part of the idea. Here, freezing nuclei (fine soil particles and volcanic dust are seen as examples) provide a frame for the crystals to form. Because water vapour and ice have different pressure characteristics we find that the water vapour is deposited on the ice crystals making them larger. Once big enough they can fall and turn to water if the conditions are right.

Rain is only one form of precipitation. If there is a strong instability in the area then a thunderstorm could develop. A thunderstorm is composed of a number of cells each with a life cycle usually referred to as youth, maturity and old age. In the youth stage there is an updraught drawing the air mass up into the troposphere. Constant condensation supplies heat energy to continue the updraught. When it's cold enough, ice particles form and the Bergeron process means particles get larger. They don't fall because the updraught is too strong. By maturity, the updraught diminishes and the larger particles start to fall. Very heavy rain is felt on the ground. Thunder and lightning are seen in this phase (although the exact mechanism is not clear it does involve the build-up of static electricity is different parts of the cloud and between droplets). By the old-age stage, most of the energy of the storm has gone and just light rain results.

Just as not all precipitation is rain so not all condensation forms clouds. When any warm, moist airmass meets a colder surface then some form of aerosol can form. At ground level this would mean fog and mist if warm enough and frost if the surface were freezing whilst dew is liquid rather than solid on the surface. Finally, precipitation is studied not just in terms of clouds and air masses but in terms of its characteristics and types. The characteristics of precipitation would include data on the nature of its intensity, extent and frequency. Collection of such data helps us classify the nature of precipitation in various parts of the world. There are three types of precipitation depending upon the way it's formed. Convective rainfall is produced when an airmass is heated and it subsequently rises. Cyclonic rainfall is associated with the kind of frontal weather in the UK. Orographic rainfall is associated with rising air masses over mountains. Maps of these rainfall types demonstrate that each one has its own key area i.e. each area in the world is supplied with precipitation in its own specific way.

3.4.4   How does it vary?

Variation can take a number of forms: through time as well as between places. Spatial variation is so great that it's better to divide this into three: global patterns (world climatic regions), regional patterns (including most meteorological changes) and local (changes within a very small area such as a city).

Starting with variations through time, we find that data for time studies can come from a variety of sources including using old documents, tree-ring pollen, ice-core and ocean-core analyses. Using such data it's possible to make some approximations for Western Europe by dividing time into the six periods of the last Ice Age. The oldest is the post-glacial period. From about 10,000 years ago there was rapid warming following the last Ice Age. Sea levels rose by up to 1 m per century and by 7000 years ago the UK was an island. This was followed by the post-glacial optimum (7000–5000 years ago). Canada was still covered after Europe was free of ice. This had the effect of altering the path of depressions to the North of the UK which gave rise to warmer conditions here but led to a wetter climate in the Sahel. By 7000–4000 years ago there was a return to cooler, drier conditions which meant that in some early civilisations such as the Mesopotamians, Egyptians and Indus, irrigation was the norm. In Europe it meant a return to far wetter conditions. During the Medieval warm period it was a time of considerably warmer and drier conditions in the UK. Wheat could be grown in the north of Scotland and York was famed for its vineyards! Towards the fourteenth century however, the climate became less favourable and drought conditions were common. Soon after the late Medieval droughts the climate became much cooler (little Ice Age). Even though this drop was by as little as 1°C it had implications from reduced crop ranges and yields to the regular freezing of rivers (which at least had the positive effect of allowing 'ice fairs' on the Thames in London). Social effects were common from poverty and famine to fall in life expectancy. This shows that climate changes can be both sudden and cause considerable disruption (which should be of interest today with global warming). Finally, with the twentieth century there is a time of warming particularly in the northern hemisphere although over the last few decades a cooling has been seen.

These are just the major changes over the last few thousand years. The next stage is to examine changes between places. Just as we can group small-scale ecosystems to make a global biome classification so we can link areas of similar climate patterns to make a global-scale climate map. Amongst the more common examples of *horizontal* variation are those based on plant growth (e.g. Trewartha's version in Figure 3.4.6); energy and moisture budgets using evapotranspiration (e.g. Thornthwaite and Budyko); global circulation dividing the world according to the major climatic belts taking the influence of land masses into account (e.g. Neef's

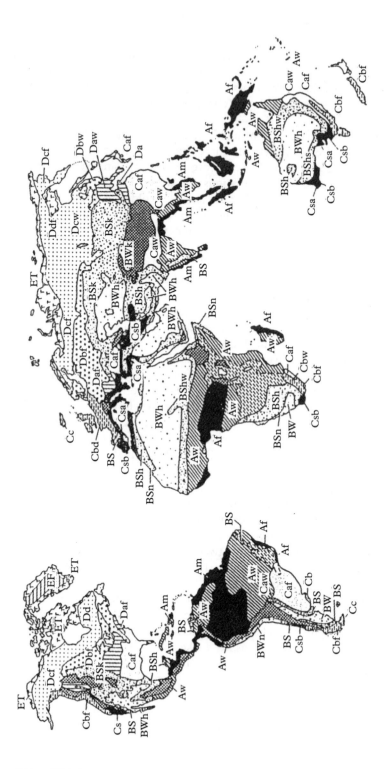

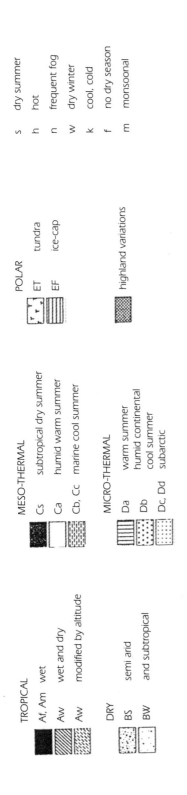

TROPICAL

Af, Am wet

Aw wet and dry

Aw modified by altitude

DRY

BS semi arid

BW and subtropical

MESO-THERMAL

Cs subtropical dry summer

Ca humid warm summer

Cb, Cc marine cool summer

MICRO-THERMAL

Da warm summer

Db humid continental cool summer

Dc, Dd subarctic

POLAR

ET tundra

EF ice-cap

highland variations

s dry summer

h hot

n frequent fog

w dry winter

k cool, cold

f no dry season

m monsoonal

Figure 3.4.6 The climates of the continents

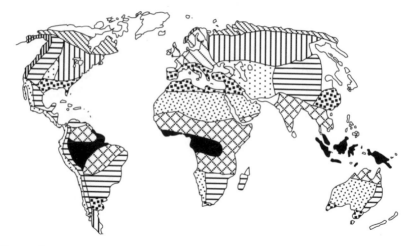

Figure 3.4.7 Classification of the world climates (after Neef)

version is seen in Figure 3.4.7 although a more comprehensive version is that of Strahler); and finally, comfort zones using measures of temperature, humidity and wind speed you can divide areas into their comfort factors for people.

There's been an interest in global *vertical* models since the seventeenth century. One example of this was the model of Hadley in 1735. Here, warm air rose over the tropics and moved northwards (for the northern hemisphere). It would become cold and start to sink. In the polar regions, cold air would flow south, be warmed and would rise. To complete this

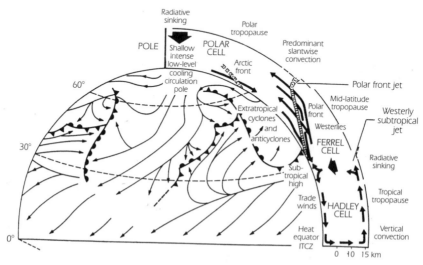

Figure 3.4.8 General circulation of the northern hemisphere

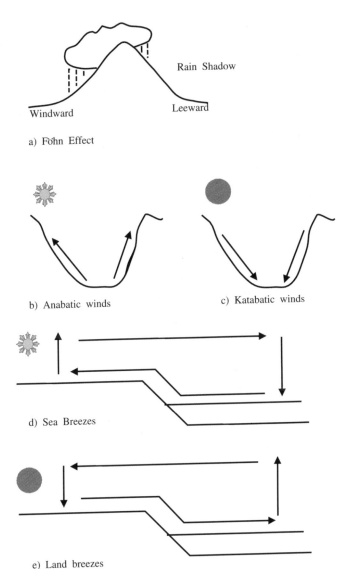

a) Föhn Effect

Rain Shadow

Windward

Leeward

b) Anabatic winds

c) Katabatic winds

d) Sea Breezes

e) Land breezes

Figure 3.4.9 Local climatic effects

picture, an indirect cell would be put in between which could be used to transfer the energy through friction with the other two. Despite this simplicity of this model parts are still accepted today (Figure 3.4.8). The tropical cell (now called the Hadley cell) still rises; the polar cell gains energy from horizontal mixing leaving the centre cell much reduced. A large temperature gradient combined with the general motion of the atmosphere leads to the creation of the jet streams in the upper troposphere. These are very high speed bands of wind (up to 480 km/hr in the winter) whose existence is critical. Originally, they were thought to be just upper atmosphere phenomena but their links with convergence and divergence means that they form a link between the ground and the upper atmosphere and so are major transporters of energy (but also pollution). These jet streams do not follow a simple path but vary three-dimensionally to create waves and troughs. Work is still being carried out to create a better model but one thing we do know is that the picture is becoming increasingly complex and we'll no longer be able to keep to a single, simple model of vertical circulation.

From the global to the regional scale concentrating on weather rather than climate (i.e. daily variation rather than general conditions). Regional variations in wind can be most important. Figure 3.4.9 shows some of the more common effects. The Fohn effect explains what happens when an air mass is forced to rise by a mountain. It cools, looses moisture and, when it sinks again, gives a rain shadow. Sea and land breezes show another illustration of that basic principle, differential heating. By day the warmer land heats up the air mass and draws an on-shore breeze onto land; by night,

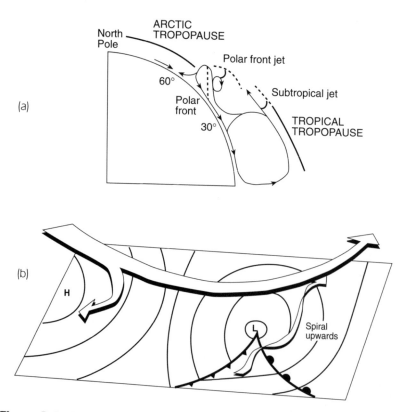

Figure 3.4.10   Jet streams

the reverse occurs. A similar situation occurs with the idea of anabatic and katabatic winds.

Depressions are common features of mid-latitude areas of the northern hemisphere (especially around the British Isles!). Pressure at the surface is not even which means that there are numerous pressure gradients resulting in the movement of air. Extremes – high pressure (or anticyclones) and low pressure (or depressions) – are both are linked to the jet streams (Figure 3.4.10). However, to get a depression instead of low pressure we need an additional feature – an airmass. In the UK there are four major airmasses each with their own characteristics of direction of origin, temperature and moisture content: polar continental (NE, cold and dry), polar maritime (NW, cold and wet), tropical maritime (SW, warm and wet) and tropical continental (SE, warm and dry). When a warm moist airmass meets a cold one under a jet stream divergence it's possible to create a depression. The sequence of events and the impact on the ground is shown in Figure 3.4.11.

The last feature is the monsoon. This is used to illustrate some of the complexities of tropical air movements. What happens when the two tropical cells meet? This area, called the inter-tropical convergence zone (or ITCZ), can give rise to some crucial weather patterns. In India this leads to monsoon conditions. In the winter, the tropical jet stream moves westwards and subsides. This, and the barrier caused by the Himalayas, causes dry air to flow east across India giving the dry season. The start of summer allows the Northern plains to heat up and cause local low pressure areas. The jet stream becomes established eastwards and brings moist air over the sub-continent. The resulting cooling of this airmass results in the strong monsoon conditions. Such wet-and-dry climates are typical of the region and although they contrast strongly with mid-latitude depressions, the basic principles involved are the same in both cases.

Local, small-scale changes are often of great importance to us in our daily lives. This will be illustrated by examining the impact of urban areas on climate. To the atmosphere, an urban area is a place where the (insolation) and albedo differ from the surrounding rural areas. With differences in energy you can get differences in pressure gradients which can affect nearly every weather feature. Three examples illustrate this point. Firstly, there's the urban 'heat island'. We use energy in towns for a variety of functions. Eventually, this degrades to heat and ends up in the air around our cities. This can build up to give a measurable difference in temperature between centre and rural outskirts of up to 6°C. Extra heat means more rapidly rising airmasses and the likelihood of instability so it's no surprise to learn that we are more likely to get thunderstorms and violent rain storms (and less snow) in our urban centres. The second case is the canyon effect. This is where the increased roughness of the surface caused by buildings (especially skyscrapers) can modify the direction and strength of the wind. The canyon effect can have other effects. On calm days it can stop pollutants from being dispersed and so they build up near the ground

DEPRESSION

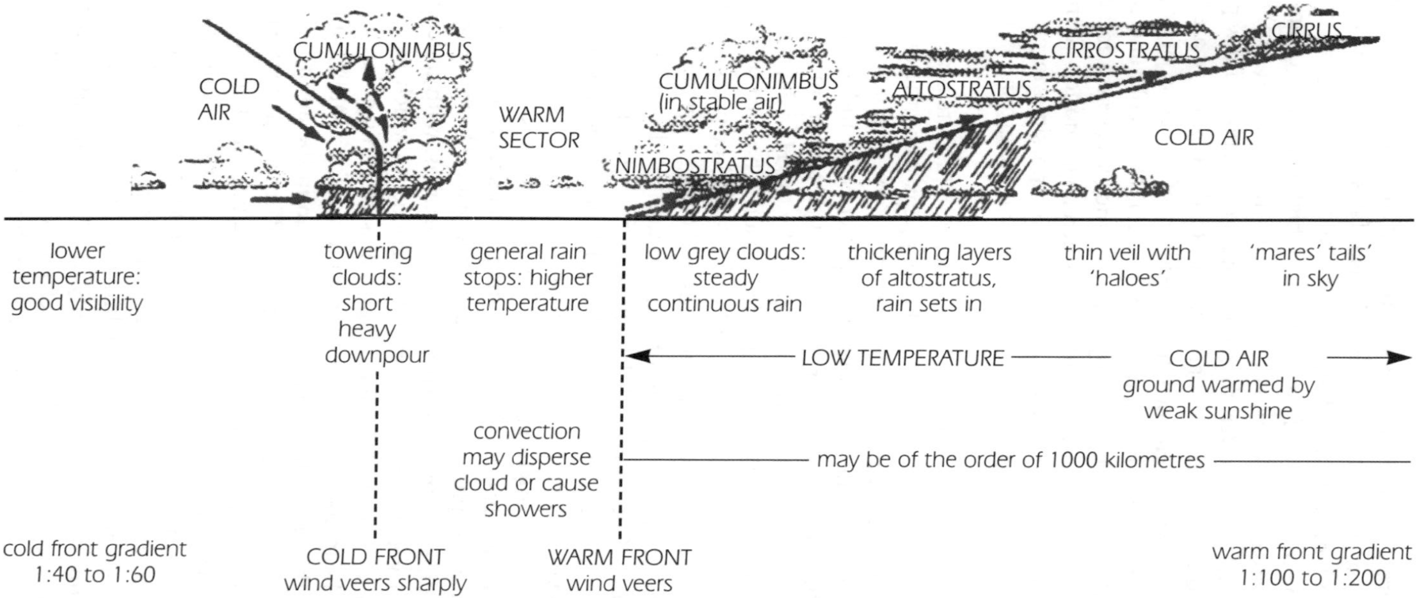

CIRRUS

CIRROSTRATUS

ALTOSTRATUS

CUMULONIMBUS (in stable air)

NIMBOSTRATUS

CUMULONIMBUS

WARM SECTOR

COLD AIR

'mares' tails' in sky

thin veil with 'haloes'

COLD AIR ground warmed by weak sunshine

thickening layers of altostratus, rain sets in

LOW TEMPERATURE

may be of the order of 1000 kilometres

low grey clouds: steady continuous rain

general rain stops: higher temperature

convection may disperse cloud or cause showers

WARM FRONT wind veers

towering clouds: short heavy downpour

COLD FRONT wind veers sharply

lower temperature: good visibility

cold front gradient 1:40 to 1:60

warm front gradient 1:100 to 1:200

COLD OCCLUSION

COLD AIR

WARM AIR

COLDER AIR at rear of depression

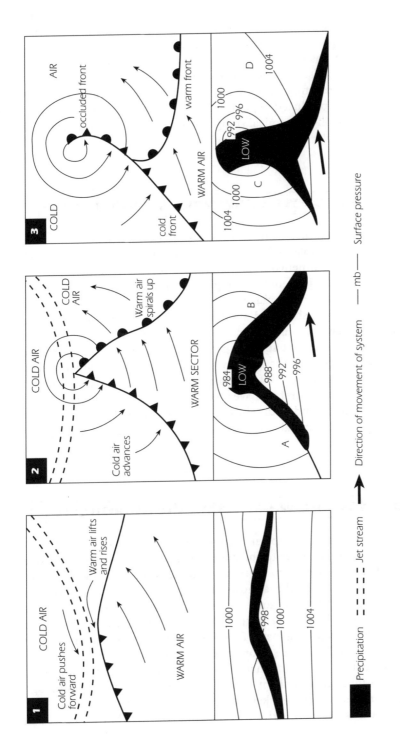

Figure 3.4.11 Sequence through a depression

Precipitation ▬ = = = = = Jet stream ↑ Direction of movement of system ─── mb ─── Surface pressure

level. Tall buildings also modify the amount of solar radiation reaching the ground which in turn affects the urban microclimate. Thirdly, there's the changing the composition of the air. Because of pollution, the amount of particles in the air has increased. These particles can be small enough to act as nuclei for moisture droplets leading to an increase in city fogs. Alternatively, they can mix with fog to give the far denser smog (smoke + fog). Gases are also produced in pollution and some of these can react with sunlight giving the photochemical smog which looks like a brown haze and which covers cites like Los Angeles so often.

3.4.5 How do we know what's happening?

Today, information comes from two main sources: international weather data gathering and public information. International efforts are co-ordinated through a global network of which the UK's Meteorological Office is a key unit. Data are gathered on a regular basis and then shared around the world for each nation to make its own maps. The usual sources were ground-based stations, ships and aircraft but today, satellites are of major importance and have revolutionised the idea of forecasting. Much of this data goes into research but some is given to the public via the synoptic charts (weather maps with a range of symbols denoting the weather at each sampling site). By using the information supplied by the Meteorological Office it's easy to interpret the weather. Newspapers regularly publish weather information and radio broadcasts offer a wider range of information from the general forecasts to the highly specialised shipping forecasts.

3.4.6 How are we affecting the atmosphere?

We have the ability to affect the atmosphere in many ways from the microscale changes in urban areas to the global changes of acid rain and ozone depletion. Before proceeding, there are two ideas worth thinking about. Firstly, any alteration in the atmosphere might have both positive and negative sides. For example, ozone depletion can lead to a net cooling if it's high level (35 km) but to net warming if it's lower level (25 km). Some studies of global warming have found some nations benefiting whilst others lose. Secondly, there's more than human input here. All natural systems fluctuate. This means that we might be calling it global warming when it's no more than a slight natural change in the system (although such ideas are becoming less likely they still shouldn't be dismissed).

Acid rain is the name given to precipitation that has a lower than normal pH (i.e. higher acidity than pH 4.5–5.0). The principal reason is human production of sulphur dioxide which has caused environmental acidity to rise by between 20% and 100% over the last 50 years. Although the problem is decreasing following agreements to limit acid emissions there's still a lot of damage reported. Attempts to reduce the problem are limited once the acidity has got into the atmosphere. Variables include atmospheric dynamics (transport, cloud type, precipitation type) and the characteristics of the receiving area (rock type, vegetation type, precipitation characteristics). Much of the problem is understood at least in basic terms with a big breakthrough coming from discovery of the link between ground level and upper-atmosphere linkages via the jet streams.

Ozone depletion became a public issue in the 1980s following the 'discovery' of reduced levels of ozone in the Antarctic regions at certain times of the year. Ozone, a highly reactive gas, is thought to be affected by industrial pollutants especially chlorofluorocarbons (CFCs) used in plastics and refrigerants. Aircraft emissions of nitrogen oxides may also play a part. Its part in reducing ultra-violet radiation from the solar spectrum is well documented and it has been feared that any reduction in this ability could increase the amount of harmful radiation reaching the Earth's surface leading to an increase in health problems (e.g. skin cancer). Nations close to the Antarctic e.g. Australia are undergoing education programmes to ensure that people use sun block creams before they go out. International agreements to limit CFC production are being negotiated but the long-term effects are not known.

Finally, there's **global warming**. Recent conferences have come to the agreement that this phenomenon is likely to increase. We know that temperature changes will not be even: the poles will be affected more than the equator. This leads to a change in the pressure gradient and a rise in the severity of the weather (with more storms, hurricanes etc.). Rates and areas for precipitation will change which implies that some nations will lose out. If the loss is great enough, political forces may be involved leading to a decrease in political stability. Also, warmer climates mean less ice. The Arctic will see little change as icebergs don't occupy a greater volume than water but you could see a change in salinity and ocean currents. Antarctic melting could cause a rise in sea level but warmer seas mean an increase in volume by thermal expansion and so there would be a double effect.

All of these phenomena suggest a great change. However, it's worth noting that these conclusions are far from certain because firstly, we don't have enough data and secondly, we still don't know that much about the atmosphere in general and the part played by the oceans in particular. So, while we are still studying the atmosphere it makes sense to keep the basic principles in mind for informed geographical discussion.

Questions

1. Describe the functioning of the atmospheric system.
2. Outline the advantages and limitations of the General Circulation Model.
3. To what extent are extreme phenomena such as droughts part of the natural cycle?
4. Who uses weather forecasts? How can they be made more reliable for specific users?
5. Describe the impact of acid rain on a freshwater ecosystem.
6. 'Global warming is a statistical extremity rather than a major climatic change.' Discuss.

3.4.6 References

Barry, R.G. and Chorley, R.J. (1992) *Atmosphere, Weather and Climate*, sixth edition. Routledge.

Money, D.C. (1988) *Climate and Environmental Systems*. Collins Educational.

Musk, L.F. (1988) *Weather Systems*. Cambridge University Press.

Transport

Energy

Economics

Agriculture

Settlement

Tourism

Human

Geography

Population

Pollution

Ecosystems

Earth processes

Biological
Conservation

Geomorphology

Biophysical Geography

Biogeography

Soils

Oceanography

Meteorology

Subject overview: ecosystems

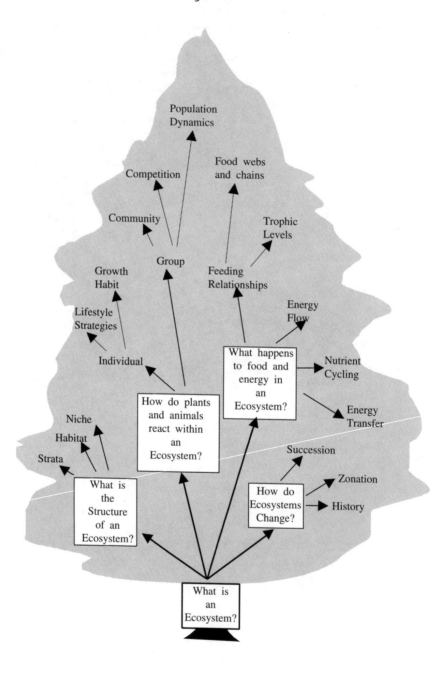

1. An ecosystem is a unique set of plants and animals and the corresponding physical conditions in which they live.
2. All components in an ecosystem interact with one another.
3. Alter any one component in an ecosystem and others will also be altered. If this is extreme enough a new ecosystem will develop.
4. Habitat is the address where the niche is the job: strata help to increase the availability of diversity of both.
5. r- and K- strategists are two ways of ensuring your species' survival.
6. Competition for resources happens between and within species.
7. The sum of all changes in the size and distribution of species is reflected in the population dynamics.
8. Food chains and webs are the fundamental ways in which relationships can be illustrated; pyramids show quantitative relationships whilst energy flow diagrams demonstrate the efficiency of links in the system.
9. Nutrient cycles are vital to our understanding of ecosystem maintenance.
10. Zonation is a change through space: succession, a change through time.

3.5.1 Introduction

Increasingly, people are concerned about the natural environment and the way in which humans impact upon it. Changes can range from the small scale like the filling-in of a village pond to large-scale phenomena like desertification. Each change creates an effect in the biosphere – the realm of living things. The aim of this chapter is to outline the key concepts of the biosphere and to show the impact of human action.

3.5.2 What is an ecosystem?

Before examining basic concepts the most fundamental question is 'what is an ecosystem?'. Taken at its simplest form one could define an ecosystem as a 'unique set of plants and animals and the corresponding physical conditions in which they live'. Thus one can talk about a woodland ecosystem, a pond ecosystem, a grassland ecosystem etc. Perhaps the best way of looking at it is to consider the triplot diagram below:

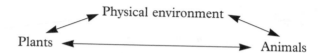

Physical environment

Plants ⟷ Animals

Note that the arrows connecting these three items are two-way. This means that each component affects the other. Plants can affect animals by being poisonous to eat and animals can affect plants by eating them but plants or animals can also affect the physical environment. For example, consider moss growing on walls and roofs. Its roots will penetrate into the brick and tile and cause small particles to be levered out (by pressure from the root systems). If this goes on for long enough the surface can become broken into particles. The moss alters the solid surface and produces particles. Although this is a human-made surface, the same process on rock is fundamental to making simple soils.

> **Tip**
>
> The important thing to remember is that all three factors are closely bound together – alter any one and you alter the others.

Having considered the basis of the ecosystem the next stage is to introduce three important issues: **boundary problems, scale** and **systems**. One of the problems of talking about ecosystems is that no one will agree on precisely what area is occupied by any given ecosystem. Take a pond – where does the pond actually end and the dry land begin? It depends upon what is considered to be a freshwater plant as distinct from a terrestrial one as well as the time of year, amount of water in the pond etc. The difficulties of drawing the line are referred to as boundary problems because we can't say precisely where one ends and another begins. In reality of course there is no such sharp divide but a gentle gradation between two systems. Scale provides another issue. It is possible to talk about a marine ecosystem which could, in theory, relate to the two-thirds of the planet covered with water. At the other end of the spectrum it is possible to describe an oak tree as an ecosystem because it contains plants and animals (in addition to the oak tree itself) and has a distinct physical environment. This could cause great confusion when trying to describe the general properties. There is some agreement nowadays with global-scale systems being called biomes and small-scale (i.e. a few square metres) systems being called micro-ecosystems.

> **Tip**
>
> The general rule is – when reading about ecosystems, check the scale. Small scale is highly detailed, large scale is very general.

Finally, the concept of the system. What is a system? Recall the idea of the system mentioned in Chapter 1.4. There should be an input (usually the physical environment in terms of water, light, temperature) and an output (growth of plants).

3.5.3 What is the structure of an ecosystem?

One of the most fundamental aspects of ecosystem study is the desire to compare one site with another not just for similar ecosystems but for different ecosystems in different parts of the world. Are there any interactions that could be seen as similar? Take the pond again and this time compare it with a woodland. At first it might seem that there are very few similarities but when one looks in more detail, the structures do contain common elements.

In terms of ecosystem 'design' there are three main concepts involved: **habitat, niche** and **stratum**. Habitat could be thought of as an address, the place where a plant or animal lives. This address would contain other plants and animals and have a set of physical conditions so in some ways it is like an ecosystem (a woodland habitat, for example). However, remember, the ecosystem is a dynamic entity; the habitat is just a physical location. A niche is the exact location occupied by an organism and could be best thought of as a job. Organisms that break down dead matter for example occupy a niche as decomposers. Finally, a stratum is a layer. Most ecosystems are not random collections of organisms but contain many well-defined areas within which many plants and animals are restricted. If this restriction is seen vertically (i.e. in layers) it's referred to as stratification. Both ponds and woodlands are stratified.

The number of habitats provided by an ecosystem is not constant: some ecosystems have more than others. This allows those ecosystems to have more species per unit area (called the **species diversity** of the area); see Figure 3.5.1. An appreciation of the **habitat diversity** as well as species diversity is crucial in our understanding of wildlife conservation. Another aspect of habitat deals with its **size** and **distribution** (often referred to as the **fragmentation** of habitat). Although the general idea is that fragmentation reduces diversity evidence is less clear cut. To test the reaction of species to fragmentation a series of plots was prepared in the Amazon rain forest. The results, shown in Figure 3.5.2 give some idea of the complexity that can arise. Note that some of the bee species have declined in smaller plots whilst some have survived almost unchanged. The question ecologists try to answer is why that should happen but all would agree that excessive fragmentation (as seen in the Dorset heathlands and chalklands) can only lead to an overall reduction in species.

Niche studies have given us some important principles: **uniqueness,**

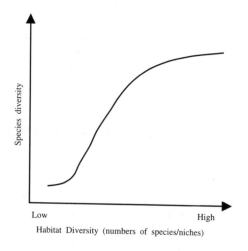

Figure 3.5.1 Relationship between species and habitat diversity

exclusivity, **coexistence** and **partitioning**. Each organism has its own unique niche. This principle holds even when one compares very similar species each within the same habitat. One study looked at birds living in pine and spruce trees. Despite similarities, detailed investigation showed that each species had its own specific feeding place (i.e. niche). What happens if two organisms share the same niche? The answer to this is shown by **'Gause's competitive exclusion principle'**. Several experiments have demonstrated this idea, perhaps the most common is with the flour beetle *Tribolium* where two very similar species appear to live in

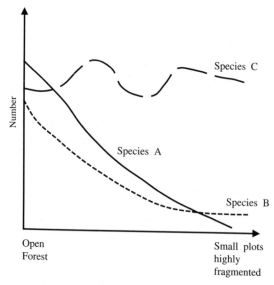

Figure 3.5.2 Plot size and bee species reaction

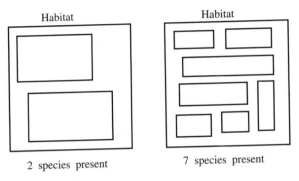

| Habitat | Habitat |

2 species present   7 species present

Figure 3.5.3 Habitat partitioning

competition with each other. Grown on their own (e.g. in a beaker of flour), both species will increase. Put them together in the same beaker and only one species will ultimately survive. Why this should occur is not known with any precision but it is thought to be related to minute differences in physical conditions giving one species the advantage. This idea of ecological 'winners' and 'losers' is called **competition**. Alternatively, species can share similar niches. Instead of competition there is niche overlap i.e. two or more closely related species have common as well as unique areas for food gathering (providing each has a food source the other doesn't). Finally, where food sharing takes place, resources are partitioned between species. There may be a case where a resource is used in common by a number of species. In this instant, experiments have shown that each species will keep to a specific range. Look at Figure 3.5.3. This is called habitat partitioning. Where two species are present then the resources are shared between them, each with a large range. Increase the number of species and the range becomes smaller.

Changes in ecosystems can be seen horizontally (i.e. by mapping). Careful study of a few ecosystems has also revealed the presence of distinct layers or strata (i.e. vertical distribution). Some, like woodlands show strata related to the growth habit of trees where there is a field layer (closest to the ground) going up to a canopy (tree tops). Some, like ponds and lakes show layers related to the changes in physical conditions. Although ponds and forests have well-documented strata there is less comment on other ecosystems although, in theory at least, they should be present.

3.5.4 How do plants and animals react within an ecosystem?

Two approaches need to be taken here: looking at the way individuals survive and the way in which groups of individuals survive.

There are two ways for an individual to dominate its part of the ecosystem: by growing in such a way that it has an advantage over its neighbours or by adopting a specific lifestyle. **Growth habit** refers to the way in which the plant develops. Heather is a good example to illustrate this: over the years it grows so as to dominate its area. The other way is to make sure that enough offspring are produced to get through to the reproducible age in the next generation. Called **lifestyle strategies** there are two extremes of a continuum. The first, seen in insects and plants for example, is to produce so many offspring that a few are bound to survive. The second pattern seen, for example, in human beings and elephants is to have very few offspring but make sure that they survive. This usually involves a great deal of parental care (ecologically speaking this means time, food and energy – all scarce resources in nature). Each strategy has a name based on the mathematical equation from which it is derived: r-strategy for the plants/insects and K-strategy for humans.

A **community** is a collection of species living in one area (usually an ecosystem). To do so they must compete with members of their own species (usually referred to as **intra-specific competition**) or members of other species (**inter-specific competition**). Assume that resources e.g. food is scarce in an ecosystem (which is usually the case). Any organism faces a number of choices: go without and starve, make sure enough can be obtained (i.e. compete with others for the limited resources) or find a food supply that no other organism wants/can utilise. The first choice is obviously no good; the second option is the most common case whilst the third suggests that a very highly specific niche can be a solution. In terms of energy saving there are definite advantages – competition costs energy and that is always in short supply.

If community describes the way in which species are grouped together (aggregate) and competition looks at how the species try to battle for survival then **population dynamics** is the sum total of their activities (a **population** is a group of the same species in the same geographical area at the same time). Populations are not static: they are subject to considerable change. In studying changes it is necessary to examine the population structure first and then to investigate relationships between populations.

The first element of population structure is **size**. This is based on the relative changes in five elements: the original population size (P), births (B), deaths (D), immigration – moving into the area (I) and emigration – moving out (E). In terms of a simple equation:

$$P(\text{year } 2) = P(\text{year } 1) + B + I - D - E$$

i.e. population growth is the starting population plus births and immigrants less deaths and emigrants.

The next most important aspect is the **age structure** (in human populations we are also concerned about the sex structure as well – the ratio of males:females). The age structure is usually represented as a **population pyramid** ($\Rightarrow$ population). The age structure is important for a number of reasons. Firstly, we can see at a glance whether the population is likely to grow or decline. Secondly, some populations have been shown to consist largely of individuals of a similar age (especially true of forests). Commercially this is useful but ecologically it can be disastrous. Recent theories about the survival of giant pandas in China have been linked to even-age stands of bamboo. Thirdly, it can be linked to the survival rates of the species.

So, change is a regular feature of all populations. This implies that there must be some mechanisms which control this process. Imagine what would happen if a population were allowed to grow unchecked. Consider this simple piece of mathematics. You have two parents who have two offspring each year. These two offspring can also breed each year after the first year. How would the population grow?

Year	0	1	2	3	4	5	6	7
Population	2	4	8	16	32	64	128	256

Such a **growth pattern** is referred to as **exponential**. If there were no restrictions the population would soon be out of control. Since this doesn't happen in nature (even with human populations contrary to some ideas) there must be something stopping it. There are several key factors. **Space** is crucial to all organisms. They need space to breed, to find enough food or just to spread. Linked to the idea of space is **territory**. Territory is that space so vital that an animal will defend it. It might contain a food supply range or be a breeding space. If an animal does not have sufficient territory then it is likely to die out. Such ideas are becoming increasingly important in nature conservation where fragmentation of habitat is seriously threatening territorial areas of key animals (especially birds of prey – why are they so affected?). **Food and water** is an obvious factor since all organisms need these to keep alive but the effect of having too little is more subtle. It might take years for a large tree to show the effects of drought for example (susceptibility to air pollution is thought to be one outcome). Lack of vital plant nutrients can retard growth. Animals with too little food might find that the reproductive potential is altered. Breeding takes a lot of energy: it's a costly business – lack of food is going to be a serious disincentive. **Herbivory** and **predation** links populations in an eat-and-be-eaten situation. Thus plants can suffer from herbivory (i.e. being eaten) to such an extent that they can die out. Likewise, predation can reduce the number of an animal species. In a famous ecological study, numbers of predator and prey were seen to fluctuate in a pattern. How can changes in number be explained? **Parasites** live on/in a host which results in the serious damage or death of that host. If this continues for long enough then the host population will suffer considerable losses. **Diseases** can reach epidemic proportions (i.e. be so common as to cause serious population loss). In the 1970s Britain suffered almost total loss of Elm trees due to Dutch Elm Disease. In medieval times the human population of Europe was devastated by the Black Death. Although parasitism and disease are rarely a major factor in nature the concept is of use to farmers and horticulturists who are turning increasingly to biological control ($\Rightarrow$ agriculture). **Weather** and **climate** can also affect populations. If conditions become extreme (such as the drought from 1991 onwards in New South Wales, Australia) then it can cause permanent harm to certain populations. **Natural disasters** are violent changes that can cause harm to natural and human populations alike and for roughly the same reasons. Take the case of the Mount St. Helen volcanic eruption in the USA. Large areas of land were covered with dust/ash and waterways were altered by the amount of debris. Habitats and homes were lost. It's not just humans that can suffer from **stress** – animals are affected also. This is one element of the concept of self-regulation: the idea, not fully understood, that animals need a 'psychological space' in order to thrive.

Add up all these factors and see that there are many ways in which a population's growth pattern can be altered. Some will affect the population as it grows in size (we call these **density-dependent** factors) other factors will operate regardless of population level (these are known as **density-independent**).

Once a population has grown as much as it can (i.e. its desire to grow has been checked by the growth factors) then it is said to have reached the **carrying capacity** of the area.

3.5.5 What happens to food and energy in an ecosystem?

What makes an ecosystem what it is? What single factor could bring together all these diverse strands together? The answer is energy. It's energy and the movement of energy that makes an ecosystem work. When we use energy we see it in terms of fossil fuels but when a plant or animal uses (or is used as) energy it is as food. Food and feeding relationships are crucial to our understanding of any ecosystem.

Feeding relationships are all about who eats what (or whom). The simplest feeding diagram is the **food chain**. This assumes that each species eats one main food source. Arrows connect species in the direction of eaten to eater (i.e. following the flow of energy), thus:

Grass $\rightarrow$ Snail $\rightarrow$ Thrush $\rightarrow$ Sparrowhawk

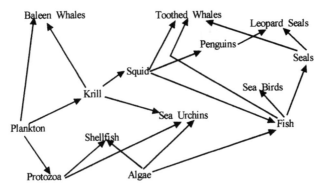

Figure 3.5.4 Simplified Antarctic food web

This food chain might work but it is too simple to describe the complexity in an ecosystem. A **food web** is a series of chains which ideally link up all the plants and animals and their feeding relationships. To get some idea of the food web look at Figure 3.5.4.

The simple food chain shown above suggests a one-way flow of energy and yet this is a system which, by definition, has some sort of feedback. So where does the feedback loop occur? All plants and animals produce waste material and eventually die. All this material (which can be called **necro-mass** or **DOM** – decaying organic matter) is broken down by a series of decomposers (such as fungi and earthworms). Some of the organisms have more than one source of food. This means that they can make some sort of choice about what they eat. Some organisms prefer one food source if it is available. This is called **food preference**. When the favoured food supply is not available, other foodstuffs will be consumed: a notion referred to as **preference** (or sometime prey) **switching**. The importance of food choice is gaining increasing acceptance. If an organism prefers one food source then it could compete directly with humans. However, if it has a more varied diet, taking what is available, then it might be far less of a problem. Until recently, seals were in this situation. Fishermen would wish to hunt them because it was thought that their food preference was commercially valuable. Now studies have shown that they consume very little valuable fish and will switch to what is available.

A food web is useful but it is no good when one tries to compare one sort of ecosystem with another. For this one turns to the idea of a feeding or **trophic level**. Rewriting the food chain shown above as a trophic diagram:

Producer → Primary Consumer → Secondary Consumer → Tertiary Consumer

Also, one can put herbivore for primary consumer, carnivore for secondary consumer with the final consumer (tertiary consumer in this case) being called the top carnivore. There are other ways to display these relationships. One could add up all the organisms involved in each trophic level

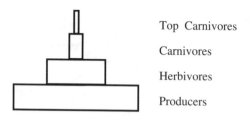

Figure 3.5.5   Pyramid of numbers

and put them into a diagram similar to Figure 3.5.5 called a **pyramid of numbers**. Sometimes this is not suitable such as when there is one oak tree being fed upon by many hundreds of insect species. One way around this problem is to estimate the mass of all the organisms involved and produce a **pyramid of biomass** (Figure 3.5.6). Finally, the food in the system could be treated as energy and an **energy flow diagram** produced (Figure 3.5.7). Here, the primary 'food source' is sunlight which enables the producers (plants) to produce energy through the process of photosynthesis. Various herbivores eat the plants and in so doing there is a loss of energy (some of the plant is not eaten: energy is needed to eat and respire). Transfer and loss continues through the system following a simple equation of **energy transfer**:

$$E_{transfer} = E_{total} - F_{uneaten} - E_{metabolism} \quad \text{(where } E = \text{energy and } F = \text{food)}$$

Tip

Although this is a qualitative idea, Lindeman, made one of the few laws of ecology by stating the efficiency of transfer between adjacent parts was about 10%.

Metabolism is the process of sustaining life and includes the energy in breathing, eating and producing waste. Farmers have another way of

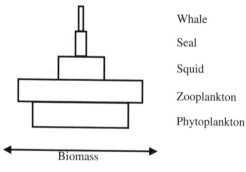

Figure 3.5.6   Pyramid of biomass

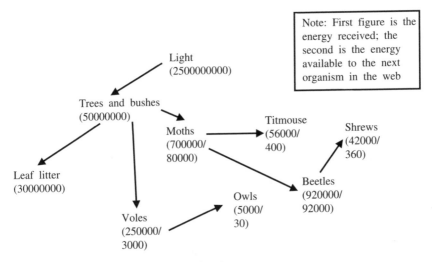

Figure 3.5.7 Energy flow in a woodland

looking at energy loss that can also be used by ecologists. Here one looks at the ratio of output of animal to input of food (called the **conversion ratio**). For example, 5 kg of cattle food will give 1 kg increase in weight of cattle (approximately). Thus it is a 5:1 ratio. Other ratios can be seen with pigs at 4:1, sheep at 3:1, chickens at 2:1.

Energy/food is not just lost through DOM , it is recycled in a process called **nutrient cycling**. In the biosphere there are a number of cycles operating all following the same basic principles. Key cycles (which are usually called **biochemical** or **biogeochemical cycles**) include carbon, nitrogen, phosphorous, potassium and sulphur. One of the more complex and about the most crucial is the nitrogen cycle. Studying it will help understand the workings of others. Examine Figure 3.5.8. Nitrogen is one of the most useful of elements, just about all organic chemicals need it but for plants and animals it is difficult to obtain. Even though 79% of the atmosphere is nitrogen gas it can only be used by organisms in another form. This form, nitrate, is produced by 'fixation' a process involving natural energy sources such as volcanoes and lightning but the main source is bacteria. These bacteria (*Rhizobium* is the main genus) produce nitrate that can be used by plants. Some plants have a symbiotic arrangement with bacteria which provide nitrate whilst living in the plant root nodules. (Such plants are called legumes – they include agricultural crops such as peas and beans.) Having fixed the nitrate it is passed through the food web until it becomes part of the DOM whence it is returned to the atmosphere via denitrifying bacteria. This means that some bacteria spend all their lives producing what other bacteria destroy! The nitrogen cycle is a good example of a natural system but its importance is such that it also has a human input in the form of artificial fertilizers. In addition to this impact, people can also create significant changes in both cycle and ecosystem by

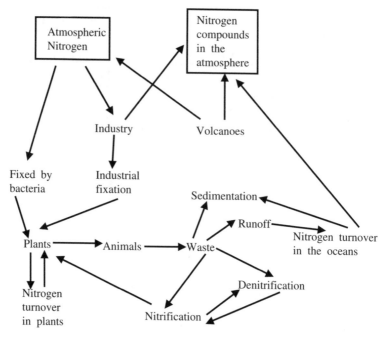

Figure 3.5.8 The nitrogen cycle

changing plant and animal communities e.g. through farming, de-forestation.

3.5.6 How do ecosystems change?

All ecosystems change. How and why they change is the subject of this section. In ecology we can study three aspects of change: change through time, which we call **succession**; change through space, which we call **zonation** and long-term global change. Of course all three aspects are of interest to ecologists but there are others who also find this information of use. For example, conservationists would look at succession and zonation as a way of understanding how species react to changing conditions. This information could then be used to ensure better reconstruction of damaged areas or help in the creation of nature reserves. A historical perspective is invaluable to climatologists who use such data to check on modern concerns such as global warming and the spread of pollution.

Imagine that a landslide has caused part of a cliff to tear away and leave a bare rock face. What will happen to it? Initially, the only plants and animals that can use such a space are those adapted to bare rock surfaces. Plants will be simple, with few roots (what is there to grow into?). Mosses and lichens are the first plants; usually called **pioneer plants**. Once their activity has broken down the surface some primitive soil might form allowing grasses and similar plants to take hold. These are the **colonisers**. If this continues then the soil will deepen and more plant species grow. Eventually, after maybe hundreds of years, some trees will take root and the area will become a woodland ecosystem. In diagrammatic form:

Bare rock → pioneer species → colonisers → early succession → late succession

There will come a time when the development will cease usually because some factor, such as climate, precludes further change. At this stage, the community is referred to as **climax vegetation**. If there has been interference in the development of the climax community e.g. through human action then it is possible that a **deflected climax** develops. Take the following case:

Bare rock → pioneers → colonisers → grassland → scrub → wood → farmland

This is roughly what happened to the English Chalk Downlands around 3000BC. The whole area was a **climatic climax** (i.e. climate halted further development) of mixed beech/ash woodland. Early settlers found the land easy to work and cleared it. This led to changes in the woodland ecosystem which meant that the grasslands that were produced by the farmers became the new climax. Because this is not the 'true' climax community it is called a deflected climax. (Some ecologists refer to each stage of such changes as a **sere** and the whole pattern becomes **seral development** with the deflected climax called a **plagiosere**.) Consider the development of a woodland (as used in the above examples). What causes it to change? Figure 3.5.9 shows that changes can be brought about by a number of factors: **allogenic** (i.e. external to the developing community) and **autogenic** (i.e. internal) factors. It is worth noting that there are also numerous factors which bring change but which do not produce a succession (e.g. see the right-hand side of the diagram).

Does every successional change bring about a specific community?

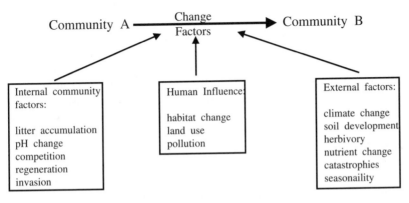

Figure 3.5.9 Change factors

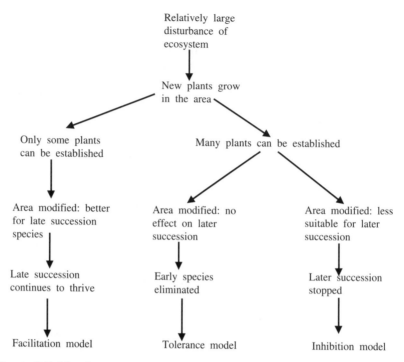

Figure 3.5.10 Succession models

This is an interesting question which has become studied increasingly over the last 25 years. The notion of the development from bare rock is accepted but what about less dramatic changes? Take the example in Figure 3.5.10. A large area becomes opened up. Three possibilities emerge. In the first case (**tolerance model**) any new invading species is eventually outcompeted by resident species. A second case allows for some of the new species to survive (**facilitation model**) whilst in the most extreme case all residents are eliminated and a new community develops (**inhibition model**). These ideas are more than just a useful study. Imagine being in charge of rehabilitating a tropical rain forest. Wouldn't you want to know which way development would go from the initial conditions?

Zonation is the change that occurs in the spatial patterns of communities. One of the best cases can be seen at the coast. When one looks at the rocks or piers one can see the distribution of various plants. Large seaweeds cling to part of the rocks but not others. Rock pools seem to have different plant communities to those on dry rocks. It might also be possible to see different areas of a rock face covered with different coloured seaweeds. All of these patterns can be referred to as zonation. How could this work? Think back to ideas on competition. The aim of the species is to survive. It can best do this in its own specific niche. A niche can have very specific habitat requirements. Alter these slightly and another species can outcompete the original one. Therefore the patterns that can be seen at the beach (or near ponds) are the visible signs of differences in physical conditions.

Succession and zonation could be assumed to be short-term changes but it is possible to study such changes over considerable timespans (when the study becomes referred to as historical ecology or **palaeoecology**). Palaeoecology has two basic tenets:

- the present is the key to the past.
- individual species do not change their habitat requirements over time.

Consider the implications of these statements. If we want to study the Ice Ages all we need is a record of some biological material at a given place. We can take a sample, record any changes and, assuming the species has not changed we can plot the climatic history. The most common example uses tree pollen. At a suitable site e.g. an anerobic bog a vertical sample is taken. This sample is divided into numerous small lengths each of which will have the pollen extracted. Study the relative amounts in each sample and make a pollen diagram showing the relative abundance of pollen. If pine is the key species it suggests colder weather. Birch suggests wetter weather and so on.

Questions —————————————————————

1. Draw a diagram of a simple ecosystem and mark on the inputs, outputs and feedback.
2. What factors could bring about changes in population sizes?
3. In a named population, describe which factors are density-dependent and which are density-independent.
4. Why do we rarely get a totally complete food web diagram?
5. What organism could have a conversion ratio approaching 1:1 i.e. where there is virtually no energy lost in metabolism?
6. What would happen if either of the two tenets (the present is the key to the past/individual species do not change habitat requirements over time) were found to be false? How could you check?

3.5.7   References

Archibold, O.W. (1994) *Ecology of World Vegetation*. Chapman and Hall.
Bradbury, I. (1991) *The Biosphere*. Belhaven Press.
Chapman, J.L. and Reiss, M.J. (1992) *Ecology: Principles and Applications*. Cambridge University Press.
Colinvaux, P. (1993) *Ecology 2*. Wiley and Sons.
Newman, E.I. (1993) *Applied Ecology*. Blackwell Scientific.

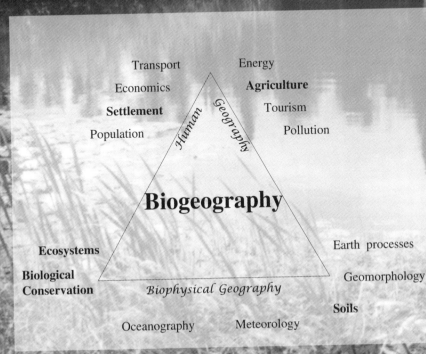

Transport   Energy

Economics   **Agriculture**

Settlement   Tourism

Population   Pollution

Human   *Geography*

Biogeography

Ecosystems   Earth processes

**Biological
Conservation**   Geomorphology

Biophysical Geography

Soils

Oceanography   Meteorology

Subject overview: biogeography

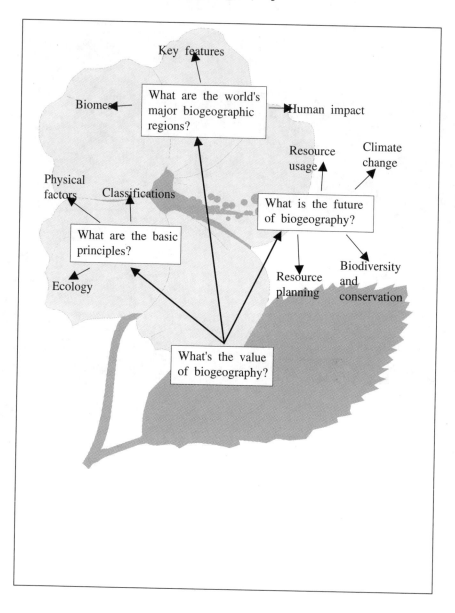

Key features

Biomes

What are the world's major biogeographic regions?

Human impact

Resource usage

Climate change

Physical factors

Classifications

What is the future of biogeography?

What are the basic principles?

Ecology

Resource planning

Biodiversity and conservation

What's the value of biogeography?

Key points

1. Biogeography is the study of plant and animal distributions on a global scale.
2. It can be understood by using a selection of ideas from ecology and the physical environment.
3. Classification is a key idea – it helps us to group smaller ecosystems into larger biomes.
4. There are 12 main biogeographical regions from pole to equator. Patterns in the northern and southern hemispheres are roughly equal but the influence of large land areas play an important role.
5. Each biome has its own ecological and physical conditions. Plants and animals respond to these through a series of adaptions.
6. With human impact increasing it is very important to understand what is happening on a global scale.

3.6.1   Introduction

Biogeography, literally the study of the distribution of plants and animal, has long been a key part of geography. Some of the earliest uses of geography have involved the search for and use of particular plant\animal regions. With the increasing use of this planet's resources it has become more important to see what's growing where. Biogeography gives us some sort of baseline from which to study.

3.6.2   What are the basic principles?

In biogeography there are three fundamental aspects: ecology, physical factors and classification. Ecology is important because we are looking at living organisms and need to know how they react. Physical factors are part of ecology but in this sense are needed to give us the global picture e.g. global climates and not just small-scale microclimates. Finally, classification is necessary: vegetational regions are the basis of the entire subject.

Ecology, the interactions of plants, animals and the physical environment is a fundamental science in biogeography because it underpins all our understanding of distributions. However, only a few of the topics are most relevant here. The first of these is **community development**. This is zonation and succession but on a grander scale. Perhaps one of the most useful approaches to this is that used by the American ecologist Eugene Odum. His idea (see Table 3.6.1) is to concentrate on those features which show how an ecosystem develops during seral changes. This can then be linked

Table 3.6.1 Changes in selected ecosystem characteristics during succession

Characteristics	Early succession	Late succession
Productivity	High	Low
Food chains/links	Linear/few	Web-like/many
Total organic matter	Low	High
Species diversity	Low	High
Community organisation	Poor	Good
Niches	Broad	Narrow
Life cycles	Short, simple	Long, complex
Life strategy	r	K
Ecosystem stability	Poor	Good

to **primary productivity**, the amount of growth per year by plants. Thus we have the areas of high productivity (rain forests and coral reefs) and areas of low productivity (deserts and mid oceans). In ecology we are used to examining plants and animals to see what sort of adaptations have been made to co-exist with the physical environment. If we are looking at changes across maybe millions of years then **evolution** becomes an important part. It's possible to find areas which have evolved completely differently. During crustal movements this land mass split. Some areas became isolated very early on and had very little input of new species e.g. Australia. Species might evolve but one fundamental should remain for each species: its **optimum range**. This is the area for which the species is best adapted and should, in theory, remain constant for a species. Such a concept can be extremely useful in plotting changes in climate (as plant distribution changes) in both time and space.

> **Tip**
>
> Optimum range is crucial to our understanding. Biogeography could be said to be the mapping of optimal ranges.

The distribution of plants and animals is usually linked to changes in climate and evolution. There may be other factors which affect the picture. One example of this could be put under the heading of **relicts** and **endemics**: species that got 'left behind' or couldn't move. There are several types of relicts. The magnolia is an example of an evolutionary relict; outcompeted by better adapted species and so only found in a few pockets of limited competition. Climatic relicts (such as the Tulip tree) have remained in pockets of favourable climate. It is thought that the biodiversity of tropical rain forests is linked to the existence of relict areas when the climate changed. Finally, there are the endemics – organisms which have become very successful but in very small areas usually quite isolated from others. To counter this there are the plants and animals that are very successful and have a very wide range. Consider the plantain, a thick-leaved weed usually found in lawns. Apart from its distribution in urban areas it has also got a global coverage. The importance of this cannot be underestimated for biogeographers. It means that plant and animal distributions are due to a variety of factors, not just the global climate but also other, evolutionary and even chance factors.

If ecological factors help explain the ways in which species interact then physical factors explain the conditions within which they operate. **Global climatic patterns** can influence global distributions. The amount of solar radiation reaching the Earth's surface varies with latitude. In some areas e.g. the Arctic the amount of radiation received can be crucial; just a small drop can mean the difference between growth and dormancy. This natural variation will, in turn, influence atmospheric patterns, the distribution of weather patterns and so the distribution of water. Of course, altitude also affects temperature and so biogeographical regions can be seen to have a three-dimensional pattern (a portion of which can be seen as Figure 3.6.1). The total amount of variation can be considerable so some climatologists have devised a global pattern to give a reasonable overview.

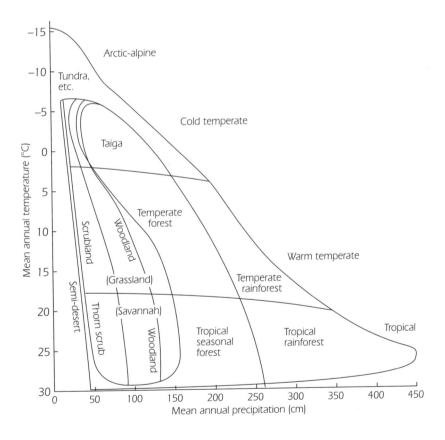

Figure 3.6.1   Distribution of biogeographical regions

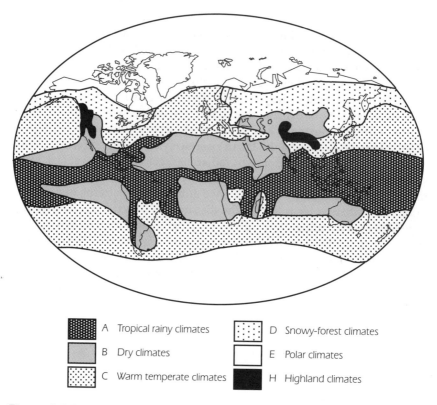

	A	Tropical rainy climates		D	Snowy-forest climates
	B	Dry climates		E	Polar climates
	C	Warm temperate climates		H	Highland climates

Figure 3.6.2   Major climatic regions of the world (after Köppen)

One such pattern is that of Köppen (see Figure 3.6.2). Although it's possible to criticise such generalisations its value lies in the fact that it works on the same scale as the biomes studied later in this chapter.

Geology is important here not so much for the distribution of rock types as the distribution of parts of the Earth's crust. Without doubt, this has been a major influence on global biogeography. Of course, as the continents have moved with their plates they have experienced changes in climatic belts (assuming that the major climatic belts e.g. tropical, temperate and polar have always been distributed in the same way). Linking both geology and climate there are the recent global changes referred to as the **Ice Age**. The recent Ice Age has had, and continues to have, a profound effect on modern plant distribution. It might be most obvious in Europe but the effects can be seen in Africa where only a few relict areas of rainforest were unaltered and so kept the biodiversity. Figure 3.6.3 shows how the last Ice Age affected the plant distribution of the Alps. Given that British vegetation was also affected less than 10,000 years ago such ideas are important to UK plant distributions. The Ice Age also shows changes through time. The **environmental gradient** illustrates how phyiscal factors changing in area can influence plant distribution. Too much (or too

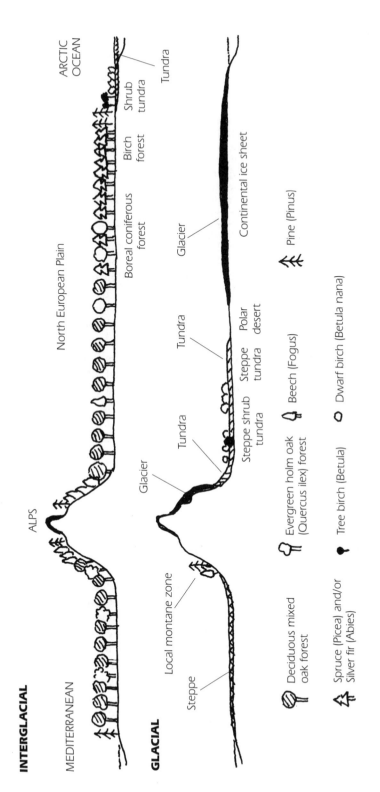

Figure 3.6.3 Affect of Ice Age on plant distribution

INTERGLACIAL

MEDITERRANEAN

ALPS

North European Plain

ARCTIC OCEAN

Tundra

Shrub tundra

Birch forest

Boreal coniferous forest

Glacier

GLACIAL

Steppe

Local montane zone

Glacier

Tundra

Steppe shrub tundra

Steppe tundra

Tundra

Polar desert

Glacier

Continental ice sheet

Deciduous mixed oak forest

Spruce (Picea) and/or Silver fir (Abies)

Evergreen holm oak (Quercus ilex) forest

Beech (Fogus)

Tree birch (Betula)

Dwarf birch (Betula nana)

Pine (Pinus)

Table 3.6.2 Soil Orders and their Characteristics

Soil Order	Characteristics
Entisol	Recent soils, very little development
Vertisol	Clayey soils typical of savannas
Inceptisol	Weakly developed; arctic soils
Aridisol	Shallow and stony, typical of desert areas
Mollisol	Rich A horizon - prairie soils
Spodosol	Acid eluviated soil, typical of conifer wood
Alfisol	Illuvial clay horizons typical of temperate forest
Ultisol	Weathered clayey soils of sub-tropical forests
Oxisol	Deep, leached soils of tropical rain forests, laterites
Histosol	Organic soils: peat, bog soil

little) water and the distribution of plants alters. A good example is the change from forest to savannah in Africa. Go (very approximately) north (or south) from the equator and the amount of rainfall diminishes and forest gives way to forest/grassland, savannah and, eventually, desert. The final physical factor is **soil**. Despite the enormous variation global distributions can be made. One common scheme is outlined in Table 3.6.2.

There have been many attempts to make global maps of vegetation and climate as an aid to drawing out general conclusions. The problem is that there are several valid answers one can come up with. Zoologists can divide the world mammal distribution into a number of regions (Figure 3.6.4a) whilst botanists can do the same with flowering plants (Figure 3.6.4b): combinations are possible (Figure 3.6.4c). The maps used in the next section to show major biogeographical regions are yet another example. In recent years there have been attempts to get a standard map agreed by all.

3.6.3 What are the world's main biogeographical regions?

In this section each region will be described in turn. Being large regions there are tremendous variations. It's not likely that any one species will be seen throughout the entire area. Any facts and figures quoted will be averages or ranges: treat these as reasonable guesses but never assume that they're the only figure. The aim is to give a brief description of each region. To link this overview with other parts of the book there will be a brief outline of human impact on each area – very few places (if any) have escaped untouched!

What are the key biogeographic regions (or **biomes**)? As with any classification, authorities differ but the 12 described below would appear to have the most support. For a global distribution of these biomes refer to Figure 3.6.5.

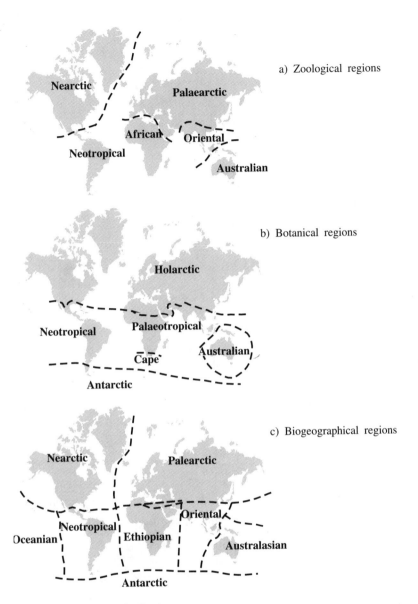

Figure 3.6.4 Global distributions

Tundra/Polar Biome. The polar regions represent some of the most marginal areas for life to colonise. The area is dominated by the lack of radiation and the low temperatures even in mid-summer – climate is a key factor. As can be seen from the map, the vast majority of this biome is found in the Eurasian/Canadian Arctic areas with some similar areas in Mongolia, along the Andean chain and in Antarctic. With low levels of solar radiation reaching the area the productivity will be low. The temper-

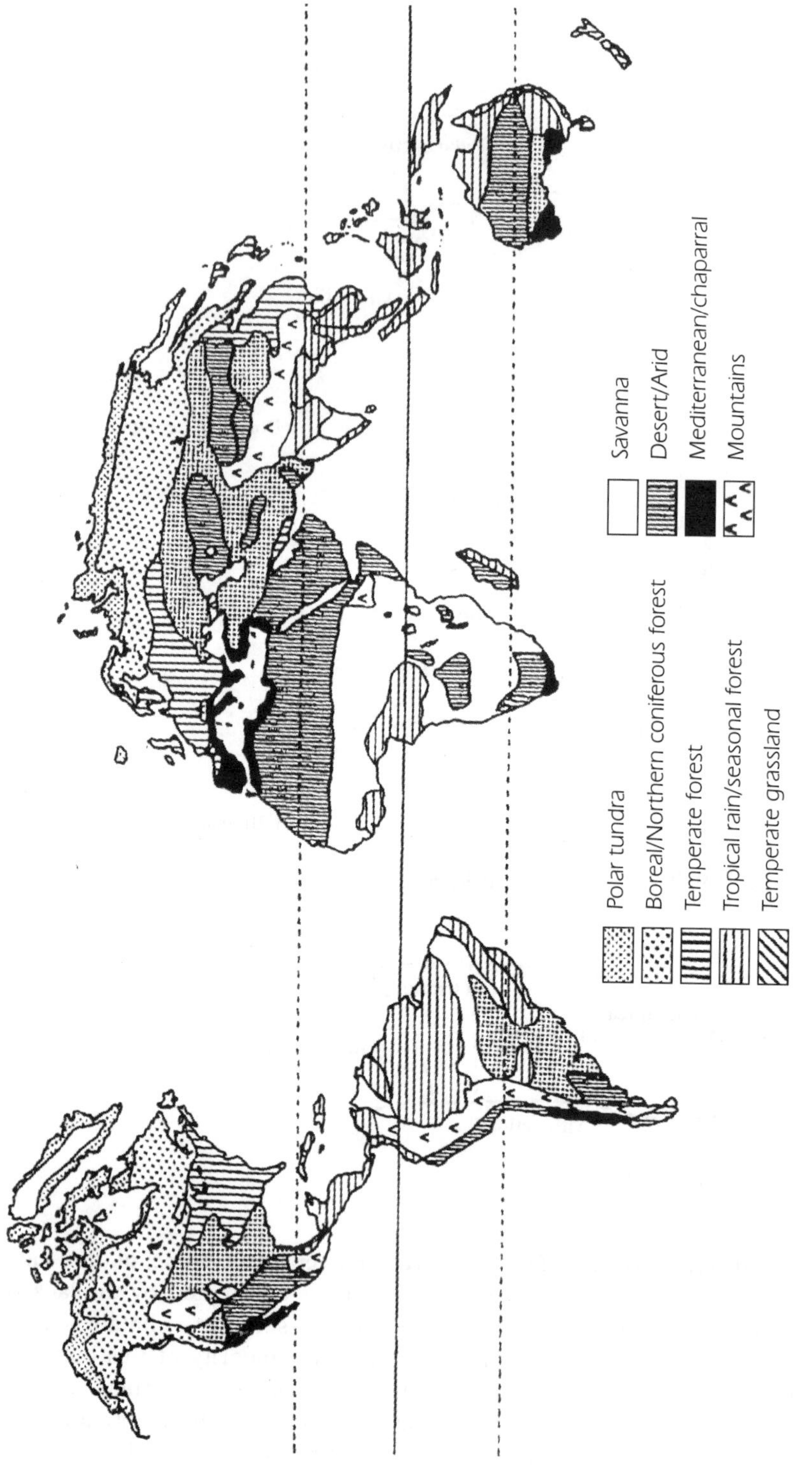

Figure 3.6.5 Global distributions

Legend:

- Polar tundra
- Boreal/Northern coniferous forest
- Temperate forest
- Tropical rain/seasonal forest
- Temperate grassland
- Savanna
- Desert/Arid
- Mediterranean/chaparral
- Mountains

atures are so low that much of the ground below 1 m is permanently frozen (the permafrost). Most of the precipitation comes from snowfall: despite the presence of so much snow the area only receives on average 250 mm per year and some areas of the Antarctic are drier than deserts. The presence of permafrost means that very little of the snow that melts infiltrates into the soil and so most meltwater will run off the slopes into the river system. This also means that the rivers have very few nutrients and this, along with temperature and daylength, comprises the limiting factors on the system. Most of the plants are mosses, lichens, sedges and dwarf shrubs such as Crowberry with a few dwarf trees away from the Poles. These are able to withstand low temperatures and freezing winds by growing close to the ground where the microclimate can be more tolerable. Snow can form a protective layer as can necromass. To go with the cold there is the very short summer growing season. Plants in this area are adapted to flowering and seeding in the few weeks available (or by growing vegetatively). The few animals that live in the region such as caribou, musk-ox and snowshoe hare have adapted to the cold with thick fur. Despite the poor conditions, the Arctic is home to many migrating bird species who live off the fruits of the short growing season.

It might be thought that human impact here has been relatively limited because of the conditions and it is certainly not highly populated. However, the area is fragile because of the marginal conditions and so any disturbance is then much more keenly felt. The native use of whale and seal species has gone on for generations but the introduction of technology created both social and environmental problems for the indigenous peoples. To this one can add the effects of resource exploitation and its impact. For example, Alaskan oil exploration has not gone unchallenged with the huge Alaskan pipeline creating considerable visual impact (as well as affecting certain migratory animals). The oil spills in the region, notably the Exxon Valdez in 1989, have drawn international attention to the problems of the region (as has commercial whaling and seal culling).

Boreal/Northern Coniferous Biome. As with many boundaries it's hard to see where polar biome ends and the boreal forest begins. In reality there's a grading from true tundra through to true boreal forest (also called the *taiga*). Boreal forest covers an immense area from the Canadian North through Scandinavia and Siberia: some 16 million km^2. The climate is milder than the polar regions but can still be harsh in winter. There is one added interest here. In the more mountainous areas, the thin atmosphere and the solar radiation in some areas can lead to a rapid heating up of the soil. At night this cools rapidly giving the area, potentially, a high daily range. As the name suggests, the dominant plant type is conifer: larch, spruce and fir being the main species. In some more southern areas deciduous trees such as birch and aspen may be seen. Soils are relatively infertile and so nutrient cycling is a limiting factor. Conifers are adapted to this to a certain extent by being able to photosynthesise throughout the year and by requiring a lower nutrient input in the first place. Fire is a major element in boreal ecology with major fires about 30 years apart. Conifers can withstand all but the fiercest fires because of their thick bark or high moisture content. Seeds are dependent upon fire: it helps breakdown the litter and allow seedlings to grow in a richer soil. Severe fires open up parts of the forest and allow the field layer to develop. Ironically, fire is critical for forest survival! Coniferous forests are best known for their timber production. About 50% of Canadian forest is suitable for logging so human impact on this scale can be considerable. Clear-felling, the defor-esting of entire slopes used to be practised but the resulting degradation of the area has meant this has stopped. Today, many logging companies are using or investigating the replanting of forests and the use of logging prac-tices that allow critical areas (such as those controlling slope erosion) to remain standing.

Temperate Forest Biome. Spreading throughout the world in the 20/40° N/S area temperate forests are a common site in Northeast America, Europe and China/Japan (with smaller but equivalent areas in Chile, Australia and New Zealand). Warm, moist summers and mild winters give rise to this deciduous forest biome. Climate is milder but is marked by seasonality. Temperature is more critical than precipitation (unlike forests towards the equator) although continentality (the climatic effects of large continental land masses) has an influence. This biome can experience a considerable variety from the 300 days per year rainfall in Chilean forests to the drier East European areas and the monsoonal Asian ones. Vegetation is primarily deciduous forest with oak, ash, hazel being good UK repre-sentatives and beeches and maples in Canada/USA. The deciduous growth habit is said to have developed as a drought protection measure. Despite the milder conditions, low temperatures are still common. Many species have developed a resistance to the formation of ice in tissues by the ability to super-cool (their freezing point is well below 0°C). One of the more common features of this type of woodland is the strong development of layers or strata. This is best seen during the European spring when the ground layer grows rapidly before the tree cover blocks out the light. (Some plants get around this problem by becoming shade tolerant.) Because of their central position in Europe and the good soils that have developed with them, temperate forests have long been subject to human impact. Early farmers and foresters cleared vast areas of Europe c.3000BC. The Weald forest was a Medieval industrial centre with charcoal making and iron production. By the end of the Medieval period the forest was much as we see it today. However, losses still occur with urban growth and farmland encroachment being the prime causes.

Temperate Grassland Biome. Temperate grasslands cover many mid-latitude areas in Eurasia and North America (where they are known as steppes and prairies, respectively). Some grasslands can also be seen in the southern hemisphere as Argentinean Pampas or South African Veld. Although the dominant vegetation is grass there are considerable variations depending on moisture: the North American prairie is subject to recurring droughts whilst the continentality of the steppes also ensures that rain-bearing winds are checked. Steppes communities are also dominated by snowfall. The southern hemisphere grasslands are usually lower lying and so climatic conditions are less extreme. The dominance of grass species has already been noted but there are variations. For example, the prairie's diversity increases the further south you go, assisted by the warmer conditions. Shorter, drought-tolerant grass species occur in the uplands. Whatever grass and herb species are present adaptation to drought is a key feature. Root structures may be stratified (to reduce competition for water) or grow to considerable depths (up to 3 m) in the search for water. Wind scorching (the damage of vegetation by constant winds) is another problem for which the reduced height of plants stops the worst excesses. In addition to climatic controls, nutrient supply is often critical which may be resolved in part by fire. Human impact has placed considerable stress on the system. Many areas are used for farming with the prairies as both best and worst example of usage. In the Britain, the use of high productivity grass species on the chalk downlands has reduced considerably the biodiversity of the area.

Savannah Biome. Savannahs are the tropical and southern hemisphere equivalent of the temperate grassland and share some of the characteristics. Although seen in large areas of South America and Australia the greatest extent is in Africa where it forms 65% of the continent's vegetation cover. The climatic pattern of savannahs tends to be more even than that of temperate grasslands. Typically, this can include a wet-and-dry yearly cycle but this tends be with a lower yearly rainfall total. It seems that seasonality rather than overall level of precipitation is the key factor here. Generally, the temperature is in excess of 18°C although it can dip as low as 4°C. Although the chief vegetation is grass there are also the mixed wood/grass communities. For example, in Australia the eucalyptus trees are found often with the savannah not least because of their ability to withstand drought. Many plants have adapted to this lack of water by growing deep root systems or developing tubers to contain water. Photosynthesis often peaks at temperatures in excess of 35°C giving plants with such a capacity an adaptive edge. As with other grassland systems, fire is a dominant factor. Human impact has been varied. Where pastoral societies have used the land then the native animals (like antelope and gazelle) have been replaced by domestic cattle. In some marginal areas this has been thought to be responsible for desertification although some researchers argue that grazing can help savannah productivity. One of the most recent moves in Africa has been the growth of tourism where the savannah offers the best views of large mammals (lions, elephants for example). This can make economic sense and some areas have developed this tourist side of the economy as a low-impact alternative. Other experiments have tried to use game meat (e.g. antelope) as a substitute for domestic cattle. Native animals are more disease resistant and can use grasslands more productively. This is not a major force as yet but it is being pursued.

Desert/Arid Biome. Deserts make up at least 25% of the land surface and many argue that this is increasing. We tend to think of the Sahara as a typical desert but the biome stretches to central Asia (as cold desert), Australia, Namibia, the USA and Chile/Peru. The idea of desert depends on your definition but most people would note the extremely low precipitation and the very wide temperature range that can be experienced (even in the Sahara the temperature can go up to 40°C and down to 4°C during the day). Drought is the key: with rainfall lacking, or sparse, or intermittent, this has to be the one limiting factor. In addition to this, cloud cover is low – down to 10% in the Sahara although up to 60% in the Asian and South American cases. If drought dominates then the flora reflect this. Giant cacti and succulents might be a major view of desert flora but this is usually a minority (often not helped by active collection of rarer species). In fact perennial grasses, shrubs and trees make up the bulk of the biomass. The key adaptation is xeromorphism which basically means that plants are structured to reduce water loss. Some plants species survive by producing growth and seeds rapidly in response to rainfall and then having the seeds dormant until the next shower. Plant cover is low (it can be as little as 8%) and so nutrient supply is a big issue: nutrients may be found only where plants already exist giving rise to 'islands of fertility'. Animals, notably carnivores, also inhabit desert regions. They have adapted to the conditions e.g. by being nocturnal, where water loss is minimised. Human impact varies according to land use. Although nomadic pastoralists move with their animals and therefore cause little lasting damage, the more static groups can overgraze an area which might lead to desertification. Some marginal desert areas have seen increased use as agricultural areas. Road construction in oil-rich areas has had a considerable impact as have poor irrigation techniques. Reclamation has been seen as a major positive step to halt desertification but it has also altered the natural biome and is not always seen as positive (especially where linked to irrigation). Demand for water by desert cities in the Middle East has lowered the water table in an area already under stress. Finally, native desert plants such as *Aloe vera* are being grown for their medicinal properties.

Tropical Forest Biome. The tropical forest biome is the most diverse of all the biomes. It consists of far more than the tropical rain forest system that is often quoted as being typical. We can distinguish three separate areas (South America, Africa and Asia) each with their own characteristics but each sharing similar properties. Climatically, as one would expect, the mean temperature is constantly in excess of 18°C and rainfall greater than 60 mm per month. Many areas see even higher rainfall due to the orographic effect. Such high temperature/rainfall figures are not universal in the tropical forest. There are considerable changes with altitude and the rain forest can give way to a simpler community with temperate species common over 2500 m. This response to changes in relief can be quite marked (called the Massenerhebung effect). The diversity of these forests is well known – up to 40% of the world's species are found here. There's no one answer to this richness but ideas have concentrated on the long history of development of the region (supported by the fossils found in the same area) and the continued existence of refugia acting as a reserve. To get this diversity in a small area means that species have to co-exist. They can do this by having large numbers of niches. Changes in light, temperature, moisture and resources open up a lot of possibilities whilst the stratification of the forest provides a vertical structure. Niche differentiation isn't the only common feature. Despite its productivity, the forest soil is usually poor: plants can't get enough nutrients from it. Various mechanisms have developed from the fast recycling of necromass by insects to the growth of large aerial roots which look as if the tree roots are growing above ground. Human impact was very limited until the twentieth century. There were numerous hunter/gatherer groups in the forest systems but they lived within the resource limitations of the areas. Today, the picture has changed so much that these forests are seen as endangered biomes with some estimates of complete loss within the next 30 years. The early farming system (*swiddening*, *gardening* or *slash-and-burn* are terms commonly used) used a small forest clearing and once fertility declined, people moved on. Population growth and declining economies in places like the Amazon have led to the opening up of the forest to large-scale agriculture (both crops and ranching), logging and mineral exploitation. Early use of forests for rubber production was always confined to a few areas in the early twentieth century and never really developed after the production of cheaper synthetics. The pressure on forests is increasing and the areas decreasing.

Mediterranean/Chapperal Biome. This is one of the smaller biomes with limited representatives on coastal areas of the Americas, Africa and Australia and the largest area being the Mediterranean itself. Originally a woodland much of the area is now a dense scrub (which goes under a variety of names: maquis, mallee and chaperral for example). The climate is seasonal with dry, high temperature conditions in summer (and subsequent fire risks) to cooler, wetter winters. The plant community of the Mediterranean gives some clue as to the response to these conditions. The original evergreen mixed forest was largely made up of oaks (such as Holm and Cork Oaks) with pines being found in the upland areas. Now, the scrub community is made up of dwarf trees (such as Myrtle) with box, juniper and gorse as common scrub plants. Plant cover is limited by the availability of moisture (which also determines plant distribution). The impact of human activity has been varied. As has happened elsewhere is

Europe the original woodland was cut for fuel and to provide pasture and space for agriculture. Continued grazing has left much of the area with poor quality foodstuff.

Wetland Biome. Other biomes might be controlled by latitude or climate, this specialised system is found where there is saturated soil. Bogs, fens, swamps and marshes are all examples of this biome whose main extent is in the boreal and tundra regions of America and Eurasia. For the soil to be saturated there must be two key conditions – a low rate of water loss through soil and an excess of precipitation over evaporation (most commonly found in Canada/Siberia). Vegetation also plays an important factor in this picture with the mosses (typically the *Sphagnum* moss) as the key group because of their water-retaining properties. In fact, mosses are the dominant vegetation although you'll also find sedge, cranberries and similar shrubs. Trees can grow in some of the drier areas. In Europe, pine, alder, birch and buckthorn are frequently seen. Wetland areas often accumulate an organic soil due to lack of decomposition of necromass. Similarly, *Sphagnum* can accumulate to form peat. Given the right conditions (i.e. anaerobic) this can become a layer thick enough to be used as fuel (e.g. Irish peat lands). Although peat is acidic and not suitable for plant growth, when areas are suitably treated (e.g. drained) they can provide very fertile agricultural areas. The Norfolk Broads has been drained since Medieval times: Irish peat areas have been vital sources of fuel with even a peat-fired power station in operation. Apart from agriculture, one controversial use is as a gardening product used in an increasing number of settings (but this is now being challenged with peat substitutes on sale). These might be UK examples but the same situation is seen wherever the peat can be used. In America, wetlands have been used to filter water pollution because of the absorptive properties of peat. Add to this an increase in tourist activity and one can see why many governments are now protecting this small biome.

Freshwater Biome. This biome is so universal it's impossible to state any particular climatic or geographic condition. Freshwater systems, which includes rivers and lakes provide vital functions (both biological and economic) but are in reality only a minute fraction of the total freshwater on the planet (1%). There is one feature that can be used to define freshwater systems – thermal stratification – which seems universal and which can be seen as the biome's one constant. Where water is still enough (e.g. lakes) a series of strata can form. These same physical variables also control the vegetation and animals found. We can distinguish a bankside (littoral) zone with rushes, reeds and a free-water (pelagic) zone with a range of floating plants (plankton, algae), and rooted plants along with a wide range of fish and freshwater invertebrates. Human impact on rivers has been considerable. They've been used for transport and of sources of food and drinking water for millennia. Recently, we've been taking water so fast that it is causing the water table to drop. Under these conditions, pollution can become a major problem from sewage, nitrate fertilizer and blue-green algae. All these examples are a danger not just to the river ecology but also human health. Some rivers are altered by the creation of artificial lakes as reservoirs: dams can also cause considerable ecological damage as has been seen in the case of the Nile in Egypt. Finally, by altering the physical characteristics of the banks (canalisation, concreting etc.) we can affect the physical responses of rivers which may lead to flooding in some areas and drought in others.

Marine Biome. If freshwater systems make up a small fraction of the water on the planet then this last biome is the largest with 70% of the Earth's surface under marine water. As with lakes there is a series of strata but these are far larger and with far greater consequences. The physical system of the seas is governed by a number of factors of which salinity, light, dissolved gases and temperature are the key ones. As with freshwater and similarly widespread biomes there is no one main ecological response to these factors. Coastal vegetation often shows considerable zonation showing that even small changes can get notable responses. Other areas, such as around the Antarctic show the effects of nutrients brought up from the ocean deep whilst coral reefs demonstrate this principle in shallower water. For centuries people have seen the oceans as challenges to travel and trade. They seemed so vast that it was unthinkable to consider that they could be damaged. Today, with increase in human activity we have a far clearer picture of what can happen. International trade has put immense pressure on coastal areas near to human populations: exploitation of resources (biological and physical) and pollution has led to some areas becoming highly damaged. Tourism has added to this in areas previously unexploited. Land-based pollution from mine waste, sewage and waste dumping has affected numerous communities. In deeper water, over-exploitation of fauna through fishing and whaling for example has given rise to global protests.

3.6.4   What is the future of biogeography?

Biogeography provides a foundation upon which we can examine changes and assess their impacts. Some of the more important changes where biogeographical knowledge can be useful are:

1 **Resource usage.** Despite giant strides in manufacturing and technology there is still an enormous demand for plant and animal material. The paperless office has yet to arrive, but writing about it consumes ever increasing amounts of woodpulp. Deforestation is a global issue

because it affects all types of woodland. Questions of the effect of such large-scale removal and the ways around these problems are the province of the biogeographer. It's not just trees that are in demand. Medicines are usually biologically based; the search for new drugs means the search for new plant and animals species that could be of help.

2 **Climate change.** The addition of a range of pollutants into the air has created a number of problems for ecosystems. 'Acid rain' has caused considerable loss of forest and river life in Scandinavia and Germany (i.e. large-scale alteration of regional biomes). The addition of other chemicals has been linked to the depletion of the 'ozone layer' with the potential for serious disruption of ecosystems affected. Other gases have been linked to the idea of global warming – that the Earth's climate is changing (some parts getting hotter and some colder, despite the name). This could cause radical changes in the distribution of biomes. One idea has the US prairies becoming a semi-desert.

3 **Resource planning.** If we are to conserve resources then we need to plan. To do this requires an understanding of the way in which biomes function. For example, how much pressure can a biome take before it is destroyed? Can we make a model of biomes that will help us to understand their workings better? Such questions are very important because they help improve our knowledge of the Earth.

--- **Questions** ---

1. Why classify? What are the advantages/disadvantages?
2. Describe in detail the distribution of a named plant/animal species. What is its optimal range? How has it adapted to this?
3. What are the problems facing biogeographers in areas such as the Amazon rain forest?

3.6.5 References

Archibold, O.W. (1995) *Ecology of World Vegetation*. Chapman and Hall.
Barbour, M.G. and Billings, W.D. (eds) (1989) *North American Terrestrial Vegetation*. Cambridge University Press.
Cox, C.B. and Moore, P.D. (1993) *Biogeography: An Ecological and Evolutionary Approach*, fifth edition. Blackwell.

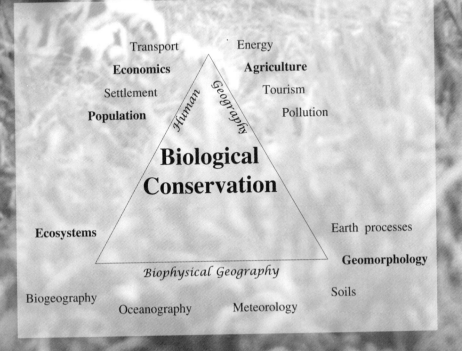

Transport   Energy

Economics   **Agriculture**

Settlement   Tourism

Population   Pollution

Human   Geography

Biological Conservation

Ecosystems   Earth processes

                                                   Geomorphology

Biophysical Geography

Biogeography   Soils

        Oceanography   Meteorology

Subject overview: biological conservation

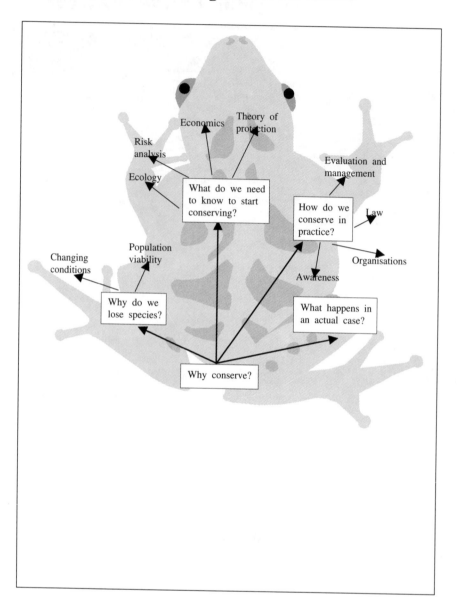

1. Conservation is the active management of resources.
2. Conservation can be justified on ecological, philosophical and economic grounds.
3. We lose species because conditions change faster than they can adapt to or because the population size falls below a critical limit.
4. Loss by human activity can be deliberate (overfishing for example) or accidental (e.g. when farming changes the countryside).
5. It is not just human pressure that causes extinctions. Even so without this it's difficult to account for population losses and gains.
6. To conserve we need to know about the ecology of the species, risk of assessment, the economics of conservation and the physical nature of the proposed site.
7. Size and shape are important factors in the nature of reserve design. Legislation and education help maintain conservation and public awareness.
8. Conservation groups play an important part in alerting us to problems.

3.7.1 Introduction

Given that some studies have shown that popular conservation organisations may well have over a million members each then it is not surprising that the topic is so popular. In the UK alone there could be a supporter of conservation in one household in five. The aim of this chapter is to introduce some of the key concepts.

3.7.2 Why conserve?

An indecent question, surely? It's obvious isn't it? Sadly, as much as conservationists would like this to be true this topic has to compete for its existence along with other geographical ideas: being popular isn't enough. In fact its very popularity has led to problems. Too many people have assumed it to be true without adequate discussion of the issues. What is conservation? Looking after plants and animals? That could include gardening!

Active management means that something has to be done to keep a species that might otherwise die out in nature. **Conservation** could be argued as being anti-nature: of keeping species against ecological pressures. **Preservation** is often linked with conservation but there is a distinct difference. Preservation is passive management: of leaving something alone or of keeping it in an unchanged state (like preserving something in formaldehyde).

It was noted above that conservation needed to be justified. Arguments against conservation range from 'species have always died out – we don't need to worry about a few more' to 'billions of people are suffering in the world: we ought to put their needs first'. What of the other side – how can we justify conservation? There are three groups of arguments: ecological, philosophical and economic. **Ecological arguments** centre around the principle of **biodiversity** (the range of species and the genetic resources they contain). Conservation promotes biodiversity by keeping a greater range of species alive. For example, conserving the tiger also involves conserving its habitat and so the plant and animal resources of a large area are secured. **Philosophical arguments** are perhaps the most difficult to justify. Many groups of people such as Buddhists see conservation as an ethical or moral issue. Put simply, because we are part of one earth then we are all of equal value. Such views can be held very deeply by people: many indigenous peoples (e.g. Aborigines, Amerindians) have survived for millennia by putting such a view at the centre of their existence. **Economic arguments** might seem to be just commercial ideas but they are increasingly powerful. Look at the rise in conservation groups and 'green' businesses thought to be worth, globally, US$600 billion by AD2000. Conserving wildlife provides numerous jobs and other economic benefits. Tourism is one of the fastest growing sectors of the economy with a turnover in the UK alone of over £9 billion (greater than agriculture and many manufacturing sectors). In countries such as Kenya and Belize the new idea of **ecotourism** is starting to grow rapidly.

3.7.3 Why do we lose species?

Whatever the reason behind it, species die out because one of two things happens. Firstly, the conditions under which the organisms live change faster than they can adapt to. Alternatively, the size of the population falls below a certain limit (usually called its **limit of viability**). Such conditions can come about by either natural or human means even though it's usually human intervention that gets all the press. Natural events such as volcanoes can alter areas so much that many of the original species die out. Although this might be disastrous in the short term, in the longer term such 'natural experiments' can offer a great deal of information about nature conservation. Even though long-term natural events can bring about the greatest changes it is the activities of humans that are the focus of conservation activity. It is convenient to divide human impact on species into two distinct areas: deliberate and accidental.

Deliberate loss of species comes about when the population is of benefit to human beings. For much of human existence, hunting has been seen as a valid way of food gathering but in some cases (e.g. whaling) it has taken the species to the edge of extinction. However, what of the loss caused by indigenous peoples hunting in their traditional ways? The argument over some groups carrying on their traditional lifestyles and commercial interests taking the same species is a very difficult one. Recently, Inuits in Northern Canada have been fighting to preserve their rights to hunt whale and seal at the same time as conservationists have tried to ban all hunting. It is a very difficult case to argue because it involves the rights of all peoples and not just conservationists. A second example is harvesting – controlled hunting of species. Fish harvesting seems like a way of controlling the amount taken but there are few checks on what individual boats do and soon a simple quota can turn into overexploitation and total loss of species. Trade is a third way of losing species.

Although deliberate use of species can lead to their extinction there is probably more pressure caused by accidental loss. For example, the change to a more mechanised farming system in the Western world has led to the reduction or loss of numerous species. Land use has changed and the habitat has been lost. Where farming has expanded it has reduced the amount of land suitable for conservation: remaining areas are just fragments of the original scattered over the countryside.

Allied to agricultural policy is habitat destruction or disruption. Too often wild areas are seen as being 'unproductive'. Habitat can be lost for building, sand and gravel extraction, industry, military and forestry uses. In the heathlands of southern England all these pressures and more have led to the decline of numerous species. Even where the area is not completely destroyed then it can still be disrupted: a wild area next to a housing estate is not the same as one in open country!

Where people have travelled to other countries they have taken their plants and animals. **Introduced species** (as they are called) have caused considerable ecological damage in many areas. Feral dogs and foxes in Australia have almost wiped out the koala in New South Wales; cats are doing the same to small marsupials (relatives of the kangaroos) in South Australia. War has been responsible for numerous losses: wildlife is just one more casualty. The Vietnam War of the 1960s and 1970s led to vast areas being defoliated and the possible extinction of valuable animal species. Pollution, particularly water pollution, is another problem especially for sensitive species. This can lead to the phenomenon of 'local extinction', the idea that one can keep a species in a nation but that it is absent in one or more regions. In the UK, dragonflies are a good example of this where a combination of pond loss and pollution have caused extinction in some areas. Finally, there is population growth. Where population pressure is greatest e.g. in parts of India and Nepal some species do not survive. What is interesting about this argument is that it is often applied to the 'Third World' whilst conveniently ignoring the fact that virtually all UK ecosystems have been altered.

3.7.4 What do we need to know to start conserving?

Tip

We know very little about why declines and extinctions occur. Often only one or two factors are taken into account. The complexity of many cases is often not explored.

Conservation is far more than just putting a species in a reserve and hoping for the best. Practical and realistic ideas for conserving a species need far more detailed knowledge. The aim of this section is to highlight some of the key ideas that need to be taken into account. These include areas of ecology, risk analysis, economics and the theory of nature reserves.

In terms of ecology, many ideas have been developed from the study of islands. These are seen as being like nature reserves because there is one area surrounded by a ('hostile') ecosystem. Continued studies in islands and other areas have demonstrated the need to appreciate a number of ecological concepts. Firstly, **population biology**. What is the species' reproductive biology? Some species have naturally small populations. It's important to know about birth and death rates, the conditions to successfully rear young and the **recruitment rate** (the number making it through to the age to reproduce). Although these ideas are for animals the same needs (e.g germination rates etc.) holds for plants. The size of the gene pool (the biological characteristics of the population) is important: too small and it could stop any large-scale restoration of the species. Most of these ideas are aimed at individuals within a population but there is also a need to be able to understand the population biology of the group. For example what is the ideal population density (the number per unit area) for your species and what is the ideal population size? Another group trait is the territoriality i.e. the space occupied by individuals in a group. Apart from setting aspects like density this information is crucial: make a reserve smaller than the target species' territory and it won't survive (which cause problems when conserving large-territory species like eagles).

The second aspect is **ecological structure**. How important is your

For the vast majority of species we just don't have enough information to make conservation a precise science.

species for the functioning of the ecosystem? Research in tropical rain forests has demonstrated the existence of **keystone species**. As the name suggests they hold the whole ecosystem structure together. If they go, the ecosystem alters completely. That's why seemingly complex systems like rainforests are so vulnerable: all those species held together by just one or two. Conservationists also recognise the value of other types of interspecies relationships such as indicator species (the ones that show the measure of health of the ecosystem) and umbrella species (they need a large area to survive but 'under' them the whole system can survive). As well as the characteristics of the species, the nature of the entire habitat is important. Very few areas are just one pure ecosystem. In reality the whole ecosystem is a mosaic of sub-ecosystems. The fragmentation of the ecosystem is also important. Too few, dispersed ecosystem-islands and conservation might fail. Remember, this doesn't have to be too much of a gap – research shows that 50 m is too much for a mouse.

The next aspect to explore is risk assessment. Seen as more common in things like insurance it's playing an increasing role in conservation. One example of this is population viability analysis. Its aim is to calculate the probability of a species becoming extinct. Take a single species like a kangaroo. Its survival depends upon rainfall and grass. If one can calculate the probability of rain (or drought) and the likelihood of grass growing then one can calculate if there is enough to sustain the kangaroo. If not, and nothing is done (i.e. conservation) then the species dies out. Whilst this is a useful tool it does have a number of shortcomings (for example, it only looks at a single species in a restricted habitat).

The third aspect is **economics**. The real trouble is that until recently, wild species and ecosystems were left out of economic calculations. They had no commercial use and therefore no commercial value. Wild areas equal free goods. How can conservation be valued when it's free? Follow this argument and anything makes better economic sense. Initially, there were ideas like the law of diminishing returns (see Figure 3.7.1) which meant that if people continued to, say, hunt whales, there would be a point where the cost of doing so was greater than the return. Keep to this and it would be possible to hunt indefinitely. Current ideas adapt the common economic idea of cost-benefit analysis: if benefits outweigh costs, go ahead. The obvious problem here is how do you put a price on a plant: what's the cost of a koala? Conservation will succeed if the profit from that (CP) is greater than the profit from development (DP) i.e. CP > DP. CP is made up of two elements: the 'use value' (UV) of the site i.e. the money to be made from using the resources and the 'non-use value' (NV) and refers to

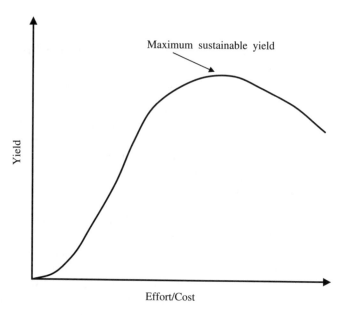

Figure 3.7.1 Law of diminishing returns

the profit to be made from keeping the site as it is (for tourism, research etc.). Thus CP = UV + NV. Conservation proceeds when UV + NV > DP.

The fourth and final aspect in this part is to examine briefly the theory of protection. Four aspects are important in making a nature reserve. Firstly, there is the focus of conservation. Is it going to be on-site (a reserve) or off-site (a zoo, botanical garden). What are the relative merits of each approach? If the on-site idea is used then is it focused on habitat or target species? Habitat conservation aims to save an entire area whilst species conservation aims at the target and conserves anything else as a byproduct of helping the target. Secondly, there is the **shape and size** (Figure 3.7.2). In terms of shape the aim is to get the ratio of perimeter:area to a minimum – a circle is ideal. The idea is to avoid what is called the edge effect ('hostile' species invading the conservation area: the 'rounder' the shape the more the centre can withstand invasion (this makes hedgerows a problem). Generally speaking, in terms of size, the bigger the better. Avoid too much fragmentation and try to get links between sites (usually called wildlife corridors). Add size and shape together and there is the interesting argument over whether it's better to have a single large or several small reserves (the SLOSS argument). Thirdly, there's the **structure** of the ecosystem. This would involve assessing a range of features to find the conservation potential of the area. Features could include naturalness – lack of human intervention; biodiversity; the presence of rare species; the ability of the area to withstand stress; typicalness – is it a good, standard example of its type; what are the surrounds like; sustainable economic value and appeal – is it likely to

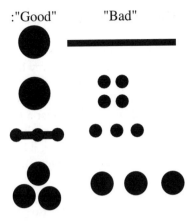

Figure 3.7.2 *Size and shape of reserves*

attract interest? Finally, there is the **construction** of the site (Figure 3.7.3). Not every area can be turned into a conservation area. Key areas need special protection. Today, many schemes use a series of areas each with decreasing conservation aims away from crucial centres:

1 **Refugia** – first seen in tropical rainforests these are areas which have never changed as far as we know. Therefore evolution and biodiversity has continued unchecked. These should be the richest areas of all.
2 **Core areas** – surrounding the refugia and having a very high conservation status.
3 **Buffer zones** – places of limited development which provide a protection for key areas and still permit the more commercial use of the area.

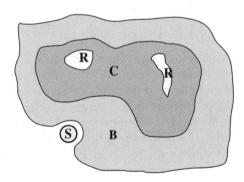

Key:
R = refugia
C = core
B = Buffer
S = settlement

Figure 3.7.3 *Refugia, core and buffer*

3.7.5 How do we conserve in practice?

It is one thing to put some theoretical ideas in place it is another to get all this into practice. Although survey after survey has come up with the idea that the majority of people support conservation there is still a vast way to go before we can say that we are conserving this planet's resources effectively. Why? Because good ideas come against a hard reality.

Not all decisions are so critical. Today, there is a growing body of ideas that can support conservation efforts. These ideas can be divided into four sections: evaluation and management, law, organisations and awareness.

When an area is suggested for conservation (whether it is big or small) the first stage is to evaluate its potential. It is no good choosing an area with a superb range of rare species if it's going to be destroyed. In carrying out this evaluation, conservationists would be looking at three aspects: general information: location, tenure (who owns it) and legal status/access (who has the rights to its resources; who can get to it?); environmental condition (see last section); and human impact. This last aspect is usually the most important for a site and would involve investigation into a number of areas including recorded history (helps to work out a strategy); intrinsic appeal (would enough people value it?); damaging operations on site/nearby (existing/potential); external influences on management and any conservation land near the proposed site

The second element is law. Evaluation might select the better sites but only law will protect them. This might be a minor problem within a nation but it is a far more difficult thing when you go international. Is it possible to get nations whose viewpoints on fundamental issues such as security and economics differ to agree on matters of wildlife? It is not easy but some remarkable progress has been made mainly through voluntary agreements called conventions. Some of the most influential are:

1   CITES – Convention on International Trade in Endangered Species of Wild Fauna and Flora. Set up in 1973 its aim is to control trade in certain animals and plants so that their conservation is not threatened and that endangered species are not driven to extinction.
2   Convention on Wetlands of International Importance especially as Waterfowl Habitat (Ramsar Convention). Wetlands are under threat everywhere. This convention, set up in 1971, aims to reduce the pressure on wetlands and co-ordinate international help.
3   World Heritage Convention. Although also covering buildings and works of art this 1972 convention aims to assist in the protection of areas of outstanding international importance.

These three examples show something of what can be done at the international level. Regional (e.g. European Union) and national (e.g. UK) authorities can also legislate to protect nature. The North Sea is a constant problem for a variety of reasons. Its high pollution levels and continued fishing has meant that the wildlife has been put under considerable pressure. A series of Resolutions has sought to bring agreement to this troubled area. In the UK, as in most nations, there are legal restrictions which seek to protect wildlife. In response to the increase in demand for recreation and the need to conserve wild areas the 1949 National Parks and Access to the Countryside Act was put in place (which gave the UK National Parks and nature reserves). By the late 1970s the pressure on wildlife led to the passing of the 1981 Wildlife and Countryside Act. It attempted to control loss of areas valuable for wildlife but its success is still subject to debate.

The third aspect concerns organisations. Despite all the work on site conservation and the legal work which keeps it in place it is probably the conservation organisations rather than anything else which have brought conservation to the attention of the public. It would be impossible to attempt to list all the groups from the small local meeting to the international giants but their effect can be profound. The best one can attempt is to produce a classification of groups:

1 **Local**. By far the most numerous, these spring up to support a local cause. Some might be single-issue groups (sometimes called pressure groups) who advocate a particular line of action or thought: others might be serious amateur groups, like field clubs, who seek to understand more about their environment.

2 **National**. There are five main groups involved at the national level. Obviously, the government (e.g. through the Department of the Environment and the Ministry for Agriculture, Fisheries and Food) is a key player as it sets the legislation and often pays other agencies to carry out the work. The second group (usually referred to as quangos) is closely allied to the government often carrying out its work. In the UK, English Nature and the Forestry Authority are good examples of this group. Learned societies (e.g. the British Ecological Society) make up the third group. Their main contribution is to bring together professionals and discuss research although many hold strong conservation views. Fourthly, there are the voluntary societies, those who exist because of their members' fees. Some, such as the Royal Society for the Protection of Birds, have memberships close to a million people and an international reputation. The fifth group comprises those pressure groups with a national presence. They are characterised by their strong views and high media profile. Friends of the Earth and Greenpeace are good examples of this type of organisation.

3 **International**. Some of the national groups have an international presence e.g. Greenpeace. Other groups are seen only at this level. Broadly they can be divided into three: governmental (collections of states with a common purpose e.g. European Community), intergovernmental (e.g. United Nations' organisations such as the United Nations

Environment Programme) and non-governmental organisations (NGOs) such as the International Union for the Conservation of Nature.

It is recognised today that no conservation will succeed without public support. To gain this support and to promote reasoned debate many governments have some form of awareness programme. Most commonly this takes two forms: general publicity/marketing and education. With tourism and interest in wildlife an increasing part of the economy it is important that nature conservation is not left behind. Awareness is a first step. The more people know about conservation the more support it is likely to get. Education is a more formal step. Many nations such as the UK and Australia have a range of courses to teach school students more about conservation.

3.7.6 What happens in an actual case?

Although there are numerous examples that can be taken, perhaps one of the easiest to appreciate is the Great Barrier Reef in Australia.

Location is fairly easy to put in general terms – it lies just off the coast of Queensland in Australia's North East. In reality, the Great Barrier Reef isn't! It's a string of reefs (about 2900) whose sizes range from 1 to 100,000 ha. In addition there are 300 cays and 250 islands. It can be 300 km from the coast in some places and nearly touching in others. The thickness of coral can vary to over 500 m whilst the age ranges from 2 to 18 million years. This is an enormously diverse set of physical environments. Although diverse it has been divided into three distinct areas (for conservation): northern (shallow water with a wide variety of reefs and mangroves); central (slightly deeper, scattered reefs) and southern (deepest part with numerous reefs).

The ecology of the area is as diverse as its geography: 400 coral species with 4000 mollusc, 350 sea urchin, 1500 fish, 6 turtle and 242 bird species not forgetting internationally important dolphin, whale and dugong areas and thousands of species of crustaceans (crabs and shrimps). In short, this is one of the most diverse ecosystems on this planet. Coral might look like a plant but it is really a vast collection of small, primitive animals called polyps. Many coral species secrete a hard shell which will build up over successive generations to make the reef structure which is so easy to recognise. To do this requires a very specific set of physical conditions. The temperature must be above 18°C (below that the coral won't grow fast enough and may even be outcompeted). Salinity (saltiness) may also be crucial. If it's not close to that found in open ocean then coral won't grow (which means it is not going to be found around freshwater outlets). The water should be clear – excessive sediment can be lethal. Like rainforests, reefs are actually in nutrient-poor surroundings: their luxuriant appearance comes from very efficient recycling of nutrients. This means that nutrients are limiting factors – any additional nutrient can disturb or destroy the natural balance.

The Great Barrier Reef is a World Heritage site but it is still under pressure not just from one but several sources of which the most important are (with their effects in brackets):

1 Beach sand for building (which increases sediment/decreases light and can also affect water flow which alters the salinity).
2 Recreation and tourism. This is an internationally renowned tourist destination, known locally and the reef and rainforest coast which brings in much needed revenue to supplement falling agricultural revenues (but tourists can collect, break, fish on/in and generally disturb coral, their hotels create sewage which leads to eutrophication).
3 Harbours and shipping bring in vital supplies and take out mining products (but require dredging and can disrupt some local reefs).
4 Fishing – sport, commercial and fish farming all use the abundant reef (but you can overfish too easily).
5 Urban and agricultural development. Queensland is the fastest growing state (but this means a need for water and materials and increases pollution).
6 Boating is a popular past-time (but anchors and propellers can damage reefs).
7 Pollution is always a problem but can be more so in a system limited by nutrient input (excessive input of sewage creates eutrophic conditions, pesticides destroy areas near rivers).
8 Mining is a source of revenue. It used to be guano bird droppings as fertilizer but is now oil (while guano mining stripped islands so they have still not recovered after nearly 100 years, oil pollution is a major threat).

This is just a summary of the numerous pressures facing this area. Remember, these are not just a few ideas that can be changed at will – each aspect represents a community or an economic idea that is trying to exist in the region. The old idea would be to ban as many of these as possible. New ideas in conservation try to reduce pressures and get all people to respect the area – to ensure it stays viable (in every sense) for future generations. The key to conservation was to have one overall authority and a plan which all residents and visitors could accept. Firstly, the area was studied to determine the uses, examine the potential of the area for development and the problems faced by development. Any plan would have to address the diverse needs of agriculture, industry, population, mining, ports, 'inherited' impacts (before the plan started), tourism and accidental damage. Next it was important to take into account the needs of the local community. These were the people who had to live with the plan – if they didn't accept it then it had less chance of working. The final scheme can

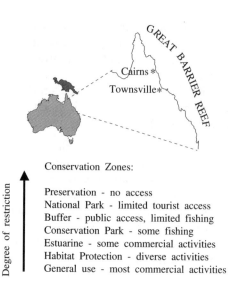

Conservation Zones:

Preservation - no access
National Park - limited tourist access
Buffer - public access, limited fishing
Conservation Park - some fishing
Estuarine - some commercial activities
Habitat Protection - diverse activities
General use - most commercial activities

Figure 3.7.4   *Great Barrier Reef Marine National Park*

be seen in Figures 3.7.4. It was considered that the most important aspect was the sustainable use of the reef. To this end the authorities decided to create a series of zones. Each zone would permit/prohibit a number of activities from the prohibited zone (where there is virtually total exclusion even for research) through a series of increasingly less protected zones to the general use zone which could be seen as a buffer zone. A crucial part of the scheme was to publicise what was happening so that everybody could see both what was needed and why. There is reason to believe that this scheme will be a useful way of controlling human impact in this unique area.

Questions

1. Name two species under threat from trade. Why are they traded?
2. Why are fragments less suitable for conservation than large areas? What could happen to a species if the fragments were smaller than their territorial needs?
3. Compare modern Ordnance Survey maps and those of the 1900s. What wildlife areas have been lost? What percentage of the area has been affected?
4. Imagine living in poverty near a rain forest. Options are starvation or illegal trading of an endangered species. What do you select? Why?
5. Using the information in the text, select a conservation site and carry out an evaluation of it? How does your result match up with the site's report. If you differ, where, why?

6. Where are the weak points in the coral's ecology (i.e. ones where human activity could easily disturb the functioning of the reef)?

3.7.7   References

Caughley, G. and Gunn, A. (1995) *Conservation Biology in Theory and Practice*. Blackwell Science.

Findtj of Nanson Institute (1993) *Green Globe Yearbook 1993*. Oxford University Press.

Given, D.R. (1995) *Principles and Practice of Plant Conservation*. Chapman and Hall

Lucas, P.H.C. (1992) *Protected Landscapes*. Chapman and Hall.

Pearce, D. and Moran, D. (1994) *The Economic Value of Biodiversity*. Earthscan.

Spellerberg, I.F. and Hardes, S.R. (1992) *Biological Conservation*. Cambridge University. Press.

Sutherland, W.J. and Hill, D.A. (1993) *Managing Habitats for Conservation*. Cambridge University Press.

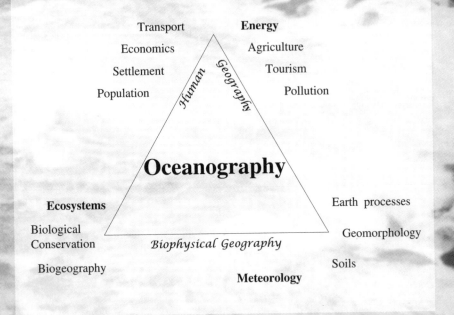

Transport
Economics
Settlement
Population

Energy
Agriculture
Tourism
Pollution

Human Geography

Oceanography

Ecosystems

Biological
Conservation

Biogeography

Earth processes

Geomorphology

Soils

Biophysical Geography

Meteorology

Subject overview: oceanography

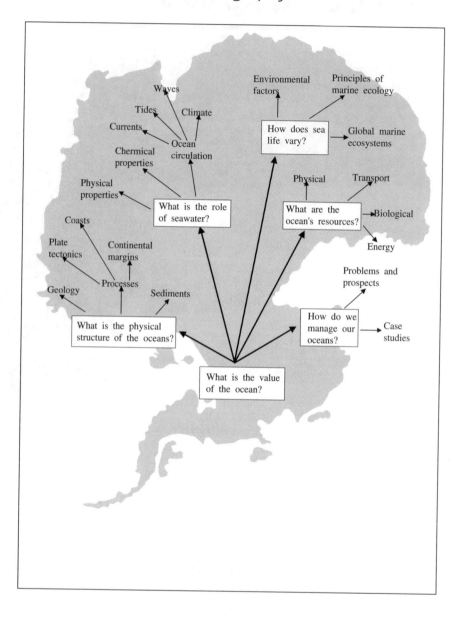

1. Water holds more heat for longer and so the oceans are a major heat sink and means of energy transfer.
2. The first 3 m of ocean contains as much energy as the atmosphere and the vast majority of water is held in oceans and so it's a major influence on the water cycle and global climate.
3. Because of the different proportions of land and ocean, southern hemisphere climate differs from northern hemisphere.
4. Ocean currents such as the Gulf Stream significantly alter climate – it is mild in the UK in winter compared with ice-locked New York.
5. Many of the processes that work in the atmosphere have equivalents in the oceans e.g. Coriolis force.
6. Human impact on the oceans has been largely negative until recently when it was seen that we could damage much of our ocean resources if we don't have management.

3.8.1 Introduction

Although they make up about 70% of the Earth's surface, oceans are one of the least studied areas. There are very few books dealing with this subject compared to other areas of geography. Where oceans are mentioned it is often as part of another topic e.g. coastal waters in geomorphology and marine life in ecology. It means that rarely is there an opportunity to study the interrelationships between marine elements which is a pity because the oceans are of vital importance in every aspect of the way in which we live. This chapter is an attempt to cover some of the most important ground: to show both the theoretical and practical use of learning more about marine areas.

3.8.2 What is the value of the oceans?

The view that because nobody owns the oceans they are free and therefore of no value is disappearing. Today, it is recognised that the oceans have considerable value. There is the physical structure of the oceans. They play a significant part in energy transfer. The Gulf Stream keeps the UK warm whilst New York is locked in winter ice. Then there are the physical resources of the oceans. Oil is an obvious choice but as we explore more we are finding a wide range of rare metals that can fetch a high price. Finally, there are the biological resources. Fishing has been a crucial economic activity for centuries. Today, we can also use other lifeforms. For example, deep water bacteria are currently being investigated for their use in pollution control. It is clear that oceans are tremendously important. The next stage is to see how they work and how we can manage them. We start by looking at the structure of ocean basins.

3.8.3 What is the physical structure of the oceans?

The ancient nature of the oceans (at least 4.5 billion years) and the idea that it seems to be the place where life originated makes the structure of our oceans much like a history book with each part giving a different section of the story. More importantly, knowledge of the structure gives us an indication of how the oceans function. There are three key areas to study: geology, sediments and processes.

Geology refers to the rocks making up the ocean floors and sides. This is important because it gives an idea of what resources are available where. It also helps us understand the way in which the earth formed. Figure 3.8.1 illustrates the various physical zones. It is the *use* we find for the geological features that is important and this is where sediment study is useful. **Sediments** come from one of three sources: weathered rock particles, decaying marine life and chemical reactions within the sea. The distribution of these three types and their proportions in any one sample can give us vital clues about the origin of these sediments and also, in some cases, about the changing environmental conditions at the surface (in particular, temperature changes). Other factors used in classification can also yield vital information: both colour and chemical composition can be used to determine the chemistry of the sediment which can again indicate origin but also give an idea of the economic potential. From these factors it is possible to recognise three broad groups of sediments: biogenous (made up largely from the hard parts of decaying organisms); lithogenous

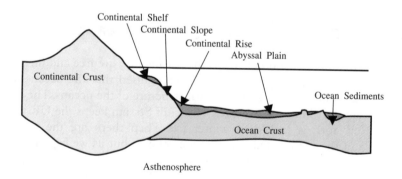

Figure 3.8.1 Ocean zones

(made from rock particles) and hydrogenous (made from chemical reactions in sea water) – see Figure 3.8.2.

The final look at the physical structure of the oceans needs to take into account the **processes** that made it all possible: plate tectonics, continental margins and coastal processes. These three operate at different scales. Plate tectonics has shaped the global ocean features. Continental margins are major regional features that exert considerable influence on us as both geographical features and as sources of much of our marine resources. Finally, the local scale of the coastal features. It is here that people actually get to interact with the marine environment so it's vitally important that we understand how it works.

3.8.4   What is the role of seawater?

The physical properties of seawater produce a series of impacts every bit as great for the marine ecosystem as the properties of air do for terrestrial systems. For example, water has a high specific heat which means it heats up and cools down slowly. For this reason alone seawater is a major moderating influence and in its way more important than the atmosphere.

> **Tip**
>
> Many of the processes that you see operating in the atmosphere operate in the oceans: it is just a question of changing the medium (i.e. from air to water).

Although the surface (i.e. horizontal) temperatures will vary according to season and distance from the equator, the vertical distribution is equally important. There are three zones: **surface layer, thermocline** where temperature declines sharply and **deep water**. These zones have been created by insolation (incoming solar radiation) at the surface and a lack of mixing at depth. This pattern is repeated for salinity and density (Figure 3.8.3). Here the surface layer is primarily restricted to open ocean where the thermocline (or **halocline** or **pycnocline**) is below it rising up only at the equator. The deep layers reach the surface only at the poles. Such a pattern is also influenced by deep ocean currents which bring nutrients from the deep and so make seemingly inhospitable polar seas teem with life.

Pressure is far greater in seawater than in air. In the deepest ocean the water can exert a pressure 1000 times greater than at the surface which also limits the lifeforms than can survive. **Light** can only penetrate so far into the ocean. Even in clear conditions only 45% of surface light can be detected at 1 m depth and this figure drops to 18% for the more cloudy

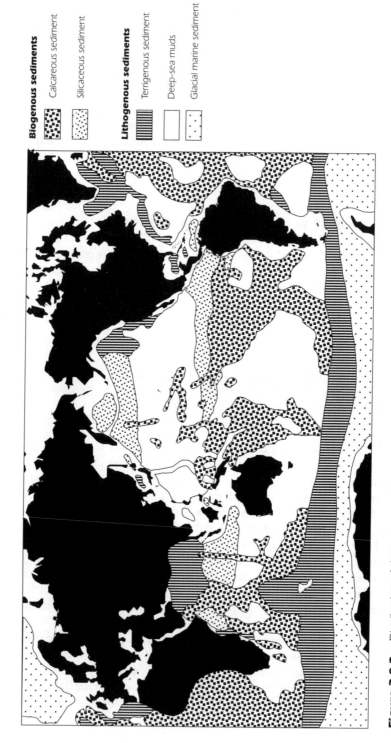

Biogenous sediments

Calcareous sediment

Siliceous sediment

Lithogenous sediments

Terrigenous sediment

Deep-sea muds

Glacial marine sediment

Figure 3.8.2   Distribution of deep-ocean sediments

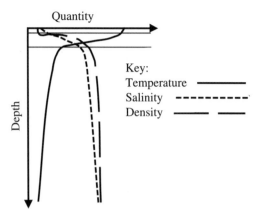

Figure 3.8.3 *Typical distribution of temperature, salinity and density in open oceans*

(turbid) coastal water. Colour will also change with depth which is often demonstrated by the zonation of different coloured algae at the coast. **Sound** is an important means of communication for both animals and humans. In the sea it means that some mammals can appear to communicate over hundreds of kilometres whilst the detection device, sonar, is used to map the ocean floor and locate important objects. The speed of sound varies with depth and pressure in much the same way that other features do.

In terms of seawater chemistry one finds that there's not just one salt but a range of salts with diverse origins. Once there, salts do not stay for ever (with the possible exception of chloride). The **residence time** is a measure of loss of salts which goes from about 100 years for aluminium to 26,000,000 years for sodium. Dissolved in seawater and influencing both chemistry and biology, oxygen and carbon dioxide are the most important gases. The concentration of both varies with depth and temperature of the water. Plant life is limited to areas where oxygen is the more available gas. Seawater chemistry acts to remove carbon dioxide as carbonate (usually calcium carbonate or chalk/limestone). Where this has happened in the past, vast layers of chalk have been deposited. In modern times, we've found out that as carbon dioxide levels increase so carbonate levels increase suggesting that the so-called greenhouse gases could be taken out of circulation faster than previously thought.

Finally we move on to the movement of seawater. Both surface and deep-sea circulations are linked. Notice that in Figure 3.8.4 each of the main oceans has a series of almost circular currents (called **gyres**) which are linked to air currents. Gyres are not single water masses. They demonstrate vertical and horizontal differences (see Figure 3.8.5 for the Gulf Stream) whilst even at the surface, different areas have different characteristics. Take the North Atlantic gyre as a case study. The Gulf Stream is the

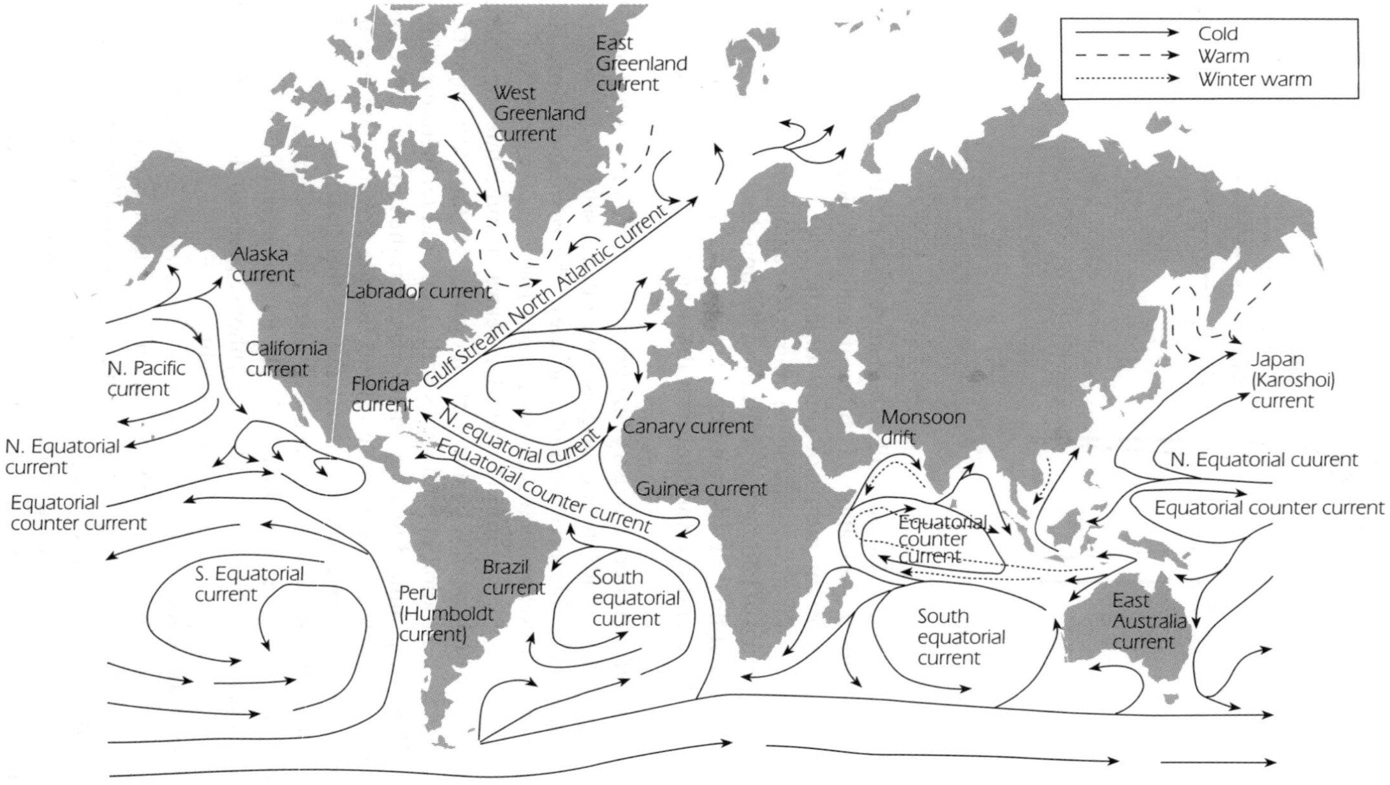

Figure 3.8.4 Circular currents and oceans

Legend:
- Cold
- Warm
- Winter warm

Labels on map:
- Japan (Karoshoi) current
- N. Equatorial cuurent
- Equatorial counter current
- East Australia current
- South equatorial current
- Monsoon drift
- Equatorial counter current
- Guinea current
- Canary current
- South equatorial cuurent
- N. equatorial current
- Equatorial counter current
- East Greenland current
- West Greenland current
- Gulf Stream North Atlantic current
- Labrador current
- Florida current
- Brazil current
- Peru (Humboldt) current
- Alaska current
- California current
- N. Pacific current
- N. Equatorial current
- Equatorial counter current
- S. Equatorial current

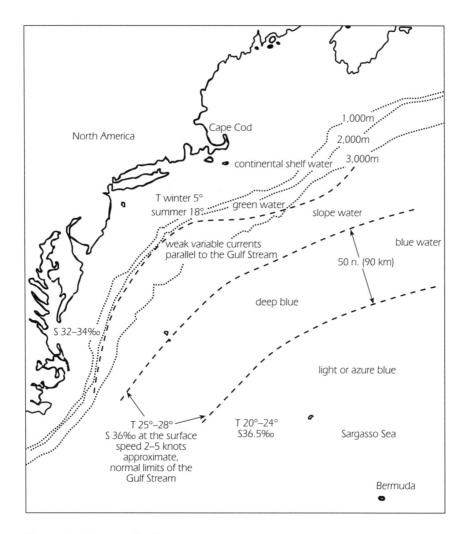

Figure 3.8.5 The Gulf Stream

Western boundary current. It is narrow (<100 km), shallow (*c.* 2km) and fast (100 km per day) and carries a tremendous volume of water each day. Eastern boundary currents carry water back towards the equator. They're generally weak and very shallow but can be extremely wide (up to 1000 km). The equatorial currents are like the trade winds – slow but almost permanent in speed and direction. They provide the return mechanism for the water.

In places such as the Pacific equatorial region there's a sub-surface undercurrent which returns a large volume of water counter to the prevailing winds. It appears that the wind will pile up water in one place and the undercurrent is compensation for this. There are divergent and convergent currents especially in equatorial and polar regions. One important case of

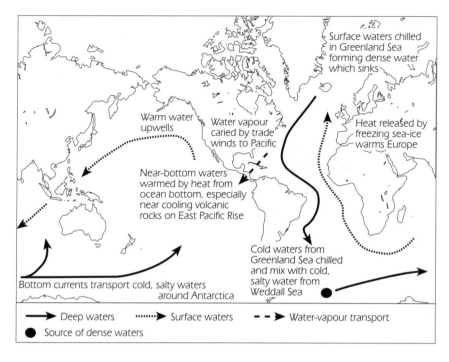

Figure 3.8.6 illustration labels:
- Surface waters chilled in Greenland Sea forming dense water which sinks
- Warm water upwells
- Water vapour caried by trade winds to Pacific
- Heat releaŝed by freezing sea-ice warms Europe
- Near-bottom waters warmed by heat from ocean bottom, especially near cooling volcanic rocks on East Pacific Rise
- Cold waters from Greenland Sea chilled and mix with cold, salty water from Weddall Sea
- Bottom currents transport cold, salty waters around Antarctica

Legend:
- ⟶ Deep waters
- ⋯⋯▶ Surface waters
- – – ▶ Water-vapour transport
- ● Source of dense waters

Figure 3.8.6 Conveyor belt theory

divergence is **upwelling**. Some upwelling is due to water meeting a land mass and some is due to the **Eckman spiral** (the changes of direction caused by the Coriolis force to sub-surface layers of the ocean). Whatever the method the result is the same. Nutrient-rich deep water is brought to the surface which gives a considerable boost to productivity. Finally, there are coastal currents, a mixture of surface current, wind and the nearness of land. These tend to be local phenomena.

Because of its importance, the El Nino effect deserves special mention. An irregular phenomenon occurring every three to seven years it has considerable oceanic and atmospheric consequences. The main cause is an atmospheric phenomenon known as the Southern Oscillation. Usually, the trade winds create a current pushing water towards Indonesia. This makes the surface waters off South America thinner and allows the deeper, nutrient-rich water to rise to the surface (leading to a rich fish harvest). Sometimes the build-up of warm water around Indonesia builds up so much that extra-strong hurricanes are produced which travel east. This reduces the westward movement of water: a thicker-than-usual surface layer results around South America cutting off nutrient-rich water and devastating fish stocks. In addition, many areas suffer climatic extremes: droughts in Australia and storms in the Americas.

Deep currents are more difficult to study because of the problems of collecting data. Generally, it is thought that currents are formed by convec-

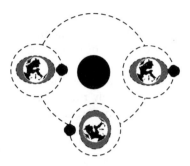

Figure 3.8.7 *Earth–Moon–Sun and tides*

tion. Warm equatorial surface waters will expand and flow north cooling all the time. Near to Greenland the water sinks and flows south, not to the equator but to the Antarctic. From here, upwelling will mean that water will flow into all oceans and start to heat up, thus repeating the cycle. This **conveyor belt theory** is a major method of transferring energy around the planet (see Figure 3.8.6).

Waves represent a movement of energy but not of water (unlike currents) caused by wind, volcanic movements or gravitational attraction. The depth of water, method of formation and the frequency of the wave can all be used to classify them: shallow water waves (near the beach, eroding), tsunamis (from underwater earthquakes), wind waves (the most common), seiche (or stationary waves), and internal waves (occurring between layers of water). A **tide** is a change in the relative height of the sea's surface. Tides are caused by the effects and relative positions of the Sun and Moon (the so-called equilibrium tidal model). As the Earth spins round the Sun it is attracted to it. The result is that the Earth bulges in the direction of the Sun. As the Earth rotates so the bulge changes position. Land bulges are too small to be noticed but water bulges (i.e. tides) are easy to spot. Adding the Moon to this simple Earth–Sun picture creates a series of tidal conditions: spring tides when Sun and Moon act together; neap tides when they are in opposition (see Figure 3.8.7). There are three types of tides: diurnal (one high and low tide per day), semidiurnal (two tides per day) and mixed (which are quite common in many parts of the world). The pattern of tides is important for our use of the sea. Not only does shipping require sufficient tides to move away from harbour but also tidal energy ($\Rightarrow$ energy) needs sufficient volume to make energy generation cost-effective.

3.8.5 How does sea life vary?

The structure of the ocean gives the boundaries and the seawater provides habitat. The oceans provide about 99.5% of all living space on Earth so there's a lot to study. To understand variations in marine ecology one must be aware of certain key factors; environmental and ecological. It is then possible to appreciate the diversity of marine life from the overview given of some ocean areas.

The properties of seawater limit life. Light is a crucial factor. Because light levels drop so rapidly it is only the surface waters (the photic zone) that can contain photosynthesising organisms. Salinity is also important causing stress in places where it changes regularly (such as estuaries). Density, pressure and depth all limit life. This is so restricting that despite the enormous biomass of the oceans it is often considered to be a biological desert. Food chains can only have so many links (about seven seems to be the maximum). Because herbivores need to be where the plant material is then community distribution is mostly limited to the photic zone. Modern distributions of marine life are linked to geological patterns. Despite the connectivity of the oceans there's very little movement between oceans. Despite these limitations, oceans are still our most productive areas. The ecological principles that produce this biomass are, of course, similar to those for land ecosystems with some important exceptions. Although there are plants in the oceans the main source of photosynthesis comes from small organisms called phytoplankton. These can only operate in the photic zone and so much of the oceans is unused. Food chains using phytoplankton are also restricted. To get around this problem there are some species that can tolerate deep water and darkness and feed on particles; other deep water species just discovered live near mid-ocean vents and use chemicals (in chemosynthesis) to gain energy. Despite the richness of some ocean areas especially near the coast in most places nutrient supply is a limiting factor. This means that wherever upwelling is found (bringing nutrients from the ocean deeps) there is an abundance of life (e.g. polar areas).

The **North Atlantic Ocean ecosystem** is one dominated by the seasons. In the depths of winter the short cold days cool the sea and surface plankton are carried down below the photic zone and reproduction stops. By the time spring comes the winds die down and more plankton make it to the surface which permits rapid growth (the so-called spring bloom). Zooplankton and other herbivores feed on the phytoplankton during the warm summers and so, towards autumn, plankton levels are reduced. There may be some nutrient mixing following storms (autumn bloom) but usually ecological activity is reduced for the winter. Thus the fisheries upon which so much European fishing depends are really seasonal in nature.

The **Sargasso Sea** is situated in the mid-Atlantic where the warm currents are surrounded by ocean currents. The thermocline is not disturbed as in the North Atlantic and so there is no upwelling of nutrients (a limiting factor). The main plant is the alga *Sargassum* which is used as food source by a range of crabs and fish. Because of the low nutrient supply growth is very slow: some plants may be hundreds of years old.

Unlike the North Atlantic, the Sargasso continues production throughout the year.

The **Antarctic**, despite climatic conditions, is one of the richest areas in the oceans. This is due to the vast upwelling of nutrients which is continuous and more than makes up for a lack of sunlight. The nutrient supply gives abundant phytoplankton. This in turn feeds an extensive food web of which whales are perhaps the best-known example.

Coral reefs do present a paradox. They have one of the highest biodiversities and a complex food web and yet they remain very fragile. One reason for this is that the coral is a keystone species – destroy that and the whole system collapses. The warm water that surrounds reefs means a permanent thermocline, no upwelling and nutrient-poor water. Many animals have symbiotic algae which help to provide the nutrients that upwelling can't supply. These measures give coral reefs the highest productivity of any marine ecosystem – up to 20 times greater than the Antarctic.

3.8.6 What are the ocean's resources?

The oceans have been a source of wealth and resources for centuries from early attempts at transport to modern oil exploration. Transportation was one of the first uses of the oceans and remains one of the most important. The majority of world trade is carried out via shipping for the advantages it gives in terms of bulk and overall cost. Shipping is also an international business with vessels owned by one nation, registered in another and crewed by yet another. Given that national economies are becoming focused on world trade this can only help to strengthen the interdependence of nations.

Physical resources are principally minerals. Despite recent problems with pollution, oil is still a major marine resource. All the recent fields have been/are marine including the North Sea, Sabah and Northwest Australia. Manganese nodules (containing manganese, cobalt, iron, nickel and copper) would have been a good source of these important minerals. Distribution is widespread but the depths (>2000 m) made economic extraction impossible. Not all resources are from modern seas. Phosphate deposits are known from fossil sea beds but in some areas such as California and Morocco they are found off the modern coast. In the nineteenth century a major source of phosphate was seabird guano found on many islands. Gold and other precious metals along with diamonds can be found in several areas where beach placers form. Finally, we are trying to use the most abundant ocean mineral – seawater. Desalination, removing the salt, is an expensive business but it can produce fresh water. Were a cheap reliable method found then it would transform many coastal regions where there is a water shortage.

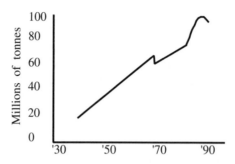

Figure 3.8.8 World fishery catch

The sea is also an energy source in its own right. Although the Romans used tide power it's only recently that there have been attempts to use it on a large scale. To modern tidal generators such as La Rance in France you can add wave power and thermal differences as potentially useful. Note that some of these sources (especially wave power) could produce a significant amount of energy at a cost similar to modern power plants but there has yet to be a serious application of the technology.

Finally, there are the biological resources whose biodiversity has been used increasingly over recent years (see Figure 3.8.8). As one can see the fish catch has increased but there is increasing evidence of overfishing with one area after another being considered as overfished. Despite such variety just a few species (herrings, cods and mullets) are caught with these examples making up 40% of the total catch. To counter this, some areas are trying aquaculture (or mariculture as it is sometimes called) which are yielding valuable stocks.

3.8.7 How do we manage our oceans?

Oceans are a paradox: they appear vast and limitless with abundant life and yet they turn out to be easily damaged and (mostly) lifeless. They provide us with economic benefits but today we see an increasing number of problems such as falling fish stocks. This must mean that management is not using these resources properly. How did we get into this situation? Too often oceans have been treated as a free good (in the economic sense) – we take rather than try to manage resources because no one owns the oceans. Apart from some attempt to have territorial boundaries out into the continental shelf no nation can claim to own a substantial part. Given that many marine ecosystems are easily stressed, there is too much use in limited productive areas and much of the ocean is a 'biological desert' it is worth looking at some of the current management strategies to find out what's happening. Four cases have been selected each of which highlights a different set of problems.

Fishing is a perfect example of the management of ocean resources as a free good. Until the twentieth century there were few limits on fishing – compared with today, catches were modest and the equipment unlikely to cause long-term damage. Even so, there were signs in the whaling industry that catches were falling and that few species were required (in this case the main catch went to smaller whales as larger ones became too few to catch profitably). From the turn of the century and especially since 1950 the world's fishing fleet changed. Boats were larger and could stay at sea longer. 'Factory ships' became the usual way of collecting deeper sea fish. Here one could catch, process and freeze stocks on an enormous scale. Throughout the North Atlantic fish stocks declined. Cod, haddock, hake and herring stocks were reduced. In the North Pacific, tuna was a key catch. However, the use of large nets trapped dolphins which were slaughtered for no good reason.

Two different areas with a common problem – how to keep up one's own fishing capacity at a time of reduced stock. What happened was that the fleets had, in trying to get all they could out of the areas, actually removed more fish than the area could produce. In other words it was overfished. These two examples show what can happen if people remove too many fish. One other example – the Peruvian anchovy fishery serves to remind us what happens if you continue to fish after a natural event has altered things. The El Nino event reduced the supply of nutrients and therefore the anchovy's food supply. However, fishing still continued which led to the collapse of the industry.

Oil pollution is a result of both ocean transport and ocean resource usage. Because most of the oil is transported before production, many of the oil refineries and associated industry are located on the coasts. Although we associate oil pollution with tanker accidents, that's a minority: almost as much comes from coastal urban areas as it does from tanker operations (about 33% each). When oil is released into the sea there are numerous pathways for it to travel along which has a considerable effect on the ecology of the area.

Coastal land use is more than just a few hotels on a few beaches. As trade is becoming more international so coastal areas are used to build up industry which minimises transport costs. Some nations, such as Australia, have developed such that the vast majority live within 30–40 km of the coast. This puts a serious strain on marine ecology. For example, many bays and estuaries are prime breeding grounds. They are also areas of highest productivity. Development, land infill, sewage and land reclamation all serve to damage one of the most sensitive parts of the seas.

Finally, there's the case of recreation. At present it's one of the world's fastest growing economic sectors (worth $30 billion in the USA alone). Coasts have always been popular locations and so it is no surprise to see these areas are under considerable strain. It's not just a question of land use there's also the issue of marine sports. Some of the best pleasure fishing areas are also in sensitive ecological areas and yet, until recently, little was being done to protect them. Australia's Great Barrier Reef is one example where such problems occur only there you have the additional problem of pollution being 'trapped' between the reef and the coast causing further difficulties.

All four examples demonstrate how use can turn into misuse. Although this presents a negative view there are some positive signs. Each of these cases can be used to show how new marine policies are protecting these resources. Many **fishing agreements** are stressing the need for yields to be regulated (since it has been obvious that the maximum sustainable yields has been exceeded). Because the open ocean is not owned by anyone, enforcement in some areas is difficult, but there are encouraging signs. Sustainable production is possible and if one could get ports rather than vessels to register catches then it might be easier to control overfishing. **Ownership** of the seas is an obstacle. In earlier times, law was settled by the use of naval power. By 1958 the United Nations recognised that there was a need for international agreement and started the first Law of the Sea Conference. The basic aim was to protect the 'common heritage' of the oceans. There were agreements on territorial limits (about 22 km) and exclusive economic zones (EEZs – 650 km from the shore) and right of passage etc. Sadly, despite years of agreement there's still some way to go. So far, only 54 nations have signed where 60 are needed to make it official policy.

Finally, we can turn to coastal management. Of the many ideas being tried here one of the more universal is the idea of zoning. In the USA, zoning is used to conserve delicate areas such as sand dunes. This means that all planning applications have to be based on what land use best suits the area rather than putting anything where people want it to go. Another use of zoning is in the Great Barrier Reef Marine National Park. Here, zoning is used to divide the coast and the sea into a series of increasingly restrictive areas. In practice this goes from general use (e.g. around ports) to highly restricted (only one or two research vessels allowed each year).

There are numerous other ways of harmonising human usage and marine areas. These cases and solutions are just a few examples to show you that it is possible to solve the problem.

Questions

1. How do you think the vertical patterns in the oceans would affect the marine ecology?
2. What is El Nino and how does it affect human activities?
3. Describe the 'conveyor-belt theory'.
4. With reference to a named marine area, describe the conservation measures being undertaken.

3.8.8 References

Barnes, R.S.K. and Hughes, R.N. (1988) *An Introduction to Marine Ecology*. Blackwell Scientific.

Gross, M.G. (1995) *Principles of Oceanography*, seventh edition. Prentice Hall.

Hansom, J.D. (1988) *Coasts*. Cambridge University Press.

Ingmanson, D.E. and Wallace, W.J. (1995) *Oceanography: An Introduction*, fifth edition. Wadsworth.

Middleton, N. (1995) *The Global Casino*. Edward Arnold.

Index

Principles and Practice Series

RESPIRATORY SUPPORT IN INTENSIVE CARE
Second edition

KEITH SYKES
Emeritus Professor,
Nuffield Department of Anaesthetics,
Honorary Fellow, Pembroke College,
University of Oxford

J D YOUNG
Clinical Reader in Anaesthetics, Nuffield Department of Anaesthetics, Oxford
Consultant Anaesthetist John Radcliffe Hospital Trust

Principles and Practice in Anaesthesia Series edited by

C E W HAHN
Professor of Anaesthetic Science, Nuffield Department of Anaesthetics,
University of Oxford, and Consultant in Clinical Measurement,
Oxford Radcliffe NHS Hospitals Trust

and

A P ADAMS
Professor of Anaesthesia, Guy's Hospital, London

© BMJ Books 1999
BMJ Books is an imprint of the BMJ Publishing Group

First published in 1995
by BMJ Publishing Group, BMA House, Tavistock Square,
London WC1H 9JR

First edition 1995
Second edition 1999

British Library Cataloguing in Publication Data

A catalogue record for this book is available from the British Library

ISBN 0-7279-1379-4

Typeset, printed and bound in Great Britain by
Latimer Trend & Company Ltd, Plymouth

Contents

Preface to first edition

When I was asked to write this book I demurred on the grounds that I had recently retired from active clinical practice. In their response the editors pointed out, firstly, that my professional life had, almost exactly, encompassed the period of most active development of mechanical ventilation and, secondly, that I had been continuously involved with the design, testing, and use of ventilators throughout that time. When I realised that few had experienced the privilege of being so intimately associated with the development of this subject I relented, and this book is the result.

It is aimed at the trainee anaesthetist, intensivist, physician, or surgeon who has the responsibility for the care of patients who require some form of respiratory support. It is didactic and stresses the principles underlying the methods used to provide respiratory support, for it is upon these principles that future developments will be based. The subject continues to advance at a rapid rate, and details of techniques and apparatus would not only have obscured the fundamental principles but would have become rapidly out of date. Nevertheless, those who have read this book should have no difficulty in approaching any of the modern microprocessor controlled machines and understanding both the purpose of the controls and the logic behind the monitoring system.

Throughout this volume I have taken the liberty of adopting a historical approach. I have done so because I believe that this adds a perspective which is often helpful when considering the possible integration of new developments into clinical practice. In the historical sections I have picked out landmark references to illustrate how one person's idea became someone else's practice. However, in the rest of the text I have restricted references to reviews, or to those papers which provide further useful information on the topic.

It would be foolhardy for anyone no longer active in the field to provide didactic statements on clinical practice. I am fortunate in having two close colleagues who agreed to read the manuscript. One is Dr L Loh, who has had a lifelong experience in the field of neurological intensive care and non-invasive methods of respiratory support, and the other is Dr J D Young, who is actively involved in intensive care research and therapy. I have also been fortunate in having two sympathetic editors, who have not only made significant contributions to the development of this subject themselves but have also been close colleagues for many years. Their comments have proved invaluable and have been incorporated into the text. However, any residual errors are entirely of my own making.

Preface to second edition

In the four years since the first edition of this book went to press there have been a few major changes in the methods used to provide respiratory support for patients in the intensive care unit. Thanks to the use of improved lung scanning techniques we now have a clearer understanding of the morphological changes associated with acute lung injury, and of the changes in the distribution of collapsed areas of lung in response to the application of positive end-expiratory pressure and the use of the prone position. There has been increasing recognition that lung damage is caused by overdistention of ventilated areas of lung and by the repeated opening and closing of collapsed areas of lung, and this has led to the widespread adoption of the so-called "open lung" techniques designed to minimise such damage. These, and attempts to improve the matching of ventilation and perfusion by the inhalation of pulmonary vasodilators such as nitric oxide and prostacyclin, have improved oxygenation in the short term but have not been shown to affect outcome. So, although the overall mortality rate from the acute respiratory distress syndrome has been falling steadily, it is difficult to pinpoint the cause of the improvement. Part of the reason for this is the extreme difficulty in carrying out prospective, randomised clinical trials in the intensive care unit environment, but the explanation might also be that there has been a gradual increase in the standard of care in intensive care units as a result of improved education, better facilities, and more advanced technology.

In the hope that this book might have contributed to the education of junior doctors working in the intensive care unit, we have produced a second edition. Dr Duncan Young, who provided much useful advice for the first edition and who is actively involved in intensive care and respiratory research, has provided the current clinical perspective and updated sections on modern techniques. The senior author has contributed the notes of caution which result from untold years of experience in this field! Trainees will find this book provides a firm physiological basis and rational guide to the increasingly sophisticated techniques used to provide respiratory support, and the historical slant will enable them to make better judgements about the value of new developments.

Conversion factors, respiratory symbols, and abbreviations used in text

SI unit conversion factors

Pressure:1 atmosphere $= 101.325$ kPa or 1013 mbar $= 760$ mmHg.
$\quad$ 1 kPa $= 7.5$ mmHg $\simeq 10$ cmH$_2$O.

Respiratory symbols

V_T or $V_E =$ tidal volume; $\dot{V}_E =$ minute volume; $\dot{V}_A =$ alveolar ventilation per minute

$f =$ breathing frequency; bpm $=$ breaths per minute

$\dot{V}_{CO_2} =$ carbon dioxide output per minute; $\dot{V}_{O_2} =$ oxygen consumption per minute

$R =$ respiratory exchange ratio $= \dot{V}_{CO_2}/\dot{V}_{O_2} = 200/250 = 0.8$ normally

BTPS $=$ body temperature and ambient pressure, saturated with water vapour

STPD $=$ standard temperature and pressure, dry (0°C, 101.325 kPa)

CaO_2, $Cc'O_2$, $C\bar{v}O_2 =$ oxygen content of arterial, end pulmonary capillary, and mixed venous blood

$FiO_2 =$ fractional concentration of oxygen in inspired gas

$PiO_2 =$ partial pressure of oxygen in inspired gas $(= FiO_2 \times$ barometric pressure $-$ water vapour pressure $= FiO_2 \times (101.33 - 6.26) = FiO_2 \times 95.07$ at 37°C

$PaO_2 =$ partial pressure of oxygen in alveolar gas

$PaCO_2 =$ partial pressure of carbon dioxide in alveolar gas

$PaO_2 =$ partial pressure of oxygen in arterial blood

$PaCO_2 =$ partial pressure of carbon dioxide in arterial blood

$P\bar{v}O_2 =$ partial pressure of oxygen in mixed venous blood

A–a$PO_2 =$ alveolar to arterial oxygen tension difference

$\dot{Q} =$ blood flow per minute

$\dot{Q}s =$ quantity of blood flowing through a shunt per minute

$\dot{Q}T =$ cardiac output per minute

$\dot{Q}s/\dot{Q}T\% =$ percentage of cardiac output flowing through a shunt

$V_D{}^{App}$, $V_D{}^{Anat}$, $V_D{}^{Alv} =$ apparatus, anatomical, and alveolar dead spaces

$V_D{}^{Phys}$ or $V_D{}^{P} =$ physiological dead space (anatomical $+$ alveolar dead space)

V_D/V_T = (physiological) dead space/tidal volume ratio (normally <0.3)
$\dot{V}/\dot{Q}$ or $\dot{V}_A/\dot{Q}$ = ventilation/perfusion ratio

Other abbreviations

ACV or A/C = assist/control ventilation
ALI = acute lung injury
AMV = assisted mechanical ventilation
APRV = airway pressure release ventilation
APV = adaptive pressure ventilation
ARDS = acute (adult) respiratory distress syndrome
ARF = acute respiratory failure
ASB = assisted breathing
ASV = adaptive support ventilation
BiPAP = bilevel positive airway pressure
BIPAP = biphasic positive airway pressure
CMV = controlled (conventional) mechanical ventilation
COPD = chronic obstructive pulmonary disease
CPAP = continuous positive airway pressure
CPPB = continuous positive pressure breathing
CPPV = continuous positive pressure ventilation
$ECCO_2R$ = extracorporeal CO_2 removal
ECLS = extracorporeal lung support
ECMO = extracorporeal membrane oxygenation
EMMV = extended mandatory minute volume
EPAP = expiratory positive airway pressure
FRC = functional residual capacity
HFCWO = high frequency chest wall oscillation
HFJV = high frequency jet ventilation
HFO = high frequency oscillation
HFPPV = high frequency positive pressure ventilation
HFV = high frequency ventilation
HME = heat and moisture exchanger
I:E = inspiration:expiration ratio
IMV = intermittent mandatory ventilation
INPV = intermittent "negative" pressure ventilation
IPAP = inspiratory positive airway pressure
IPPV = intermittent positive pressure ventilation
IPS = inspiratory pressure support
IRV = inverse ratio ventilation (inspiration longer than expiration)
MMV = mandatory minute volume
NIV = non-invasive ventilation
PC IRV = pressure-controlled inverse ratio ventilation
PCV = pressure-controlled ventilation

PEEP = positive end expiratory pressure
$PEEP_E$ = extrinsic PEEP
$PEEP_i$ = intrinsic, auto, or alveolar PEEP
PIP = peak inspiratory pressure
PRVC = pressure-regulated, volume-controlled ventilation
PS or PSV = pressure support (ventilation)
SIMV = synchronised intermittent mandatory ventilation
SV = spontaneous ventilation
VS = volume support

1 Development of techniques of respiratory support

Cyclical expansion of the lungs can be produced by generating an intermittent positive pressure in the trachea (intermittent positive pressure ventilation—IPPV) or by applying an intermittent "negative" (subatmospheric) pressure around the chest wall and abdomen (intermittent "negative" pressure ventilation—INPV). It was Vesalius who first showed that rhythmic inflation of the lungs could maintain life in open chest animals. In *De humani corporis fabrica* he wrote: "That life may in a manner of speaking be restored to the animal, an opening must be attempted in the trunk of the trachea, into which a tube of reed or cane should be put; you will then blow into this, so that the lung may rise again and the animal take in air. Indeed with a slight breath in the case of the living animal, the lung will swell to the full extent of the thoracic cavity, and the heart will become strong and exhibit a wondrous variety of motions."[1]

These experiments were repeated by Robert Hooke in London in 1664. At that time there was some doubt about the exact function of the lungs, some physicians believing that their main function was to remove body heat, while others contended that the motion of the lungs assisted the circulation of the blood. In 1667 Hooke disproved the latter hypothesis by making a large number of small holes in the surface of the lungs exposed at thoracotomy and then blowing a constant stream of air through the motionless lungs. The animal survived while the air flowed through the lungs, but it developed convulsions and died when the air flow was discontinued. Hooke also observed that the blood continued to flow through the lungs whether they were held inflated or allowed to collapse and concluded that the survival of the animal depended on the constant supply of fresh air and not the movement of the lungs.[2]

In 1776 John Hunter reported that in 1755 he had used a double bellows system to maintain life in an open chest animal.[3] These experiments led to the Royal Humane Society (founded in 1774) recommending the use of a bellows for the resuscitation of the apparently drowned. In 1827 the use of a bellows received a serious setback when Leroy reported a series of experiments showing that their use could result in severe barotrauma.[4]

1

These criticisms, and the difficulties in achieving an airtight connection between the bellows and the trachea in the drowned victim, caused the method to fall into disuse.

The lungs can be inflated by:

Applying an intermittent positive pressure to the airway—IPPV

Applying an intermittent subatmospheric or "negative" pressure to the chest wall and abdomen—INPV

- Tank ventilator: encloses the whole body with the exception of the head
- Cuirass ventilator: encloses the thorax and part of the abdomen
- Rocking bed: diaphragm moved by gravitational effects on abdominal contents
- Abdominal belt: compresses abdomen in expiration; inspiration due to passive recoil

---

Intermittent "negative" pressure ventilation

Interest then switched to ventilators that generated ventilation by producing an intermittent "negative" pressure around the chest wall and abdomen (INPV).[5,6] The body of the patient was placed in an airtight chamber with the head protruding through a neck seal and the pressure inside the chamber was intermittently reduced below atmospheric by the manual operation of a bellows or piston connected to the chamber. The earliest ventilator of this type, described by John Dalziel, a Scottish physician, in 1832, was used in attempts to resuscitate drowned or asphyxiated victims. A later American design by Alfred E Jones (1864) was used for the treatment of an extraordinary range of diseases which included "paralysis, neuralgia, seminal weakness and deafness." A number of similar machines were developed in other European countries but met with little success.

In 1924 Thunberg described the barospirator. This worked on an entirely different principle, for both the head and trunk of the subject were enclosed in a large metal chamber, gas exchange being achieved by the alternate compression and expansion of the air in the chamber and lungs. Although a number of cases of poisoning and poliomyelitis were treated in the machine in Scandinavia, the results did not encourage widespread use.[7]

The next major development occurred in 1929 when Philip Drinker of Boston described the "iron lung."[8] This was the first practical tank ventilator and variants of this machine were used to support ventilation in patients with neuromuscular paralysis due to poliomyelitis or the Guillain–Barré syndrome until the mid-1950s. The Drinker machines were very expensive and were not widely available outside the United States, but Drinker